Advanced Sciences and Technologies for Security Applications

Indexed by SCOPUS

The series Advanced Sciences and Technologies for Security Applications comprises interdisciplinary research covering the theory, foundations and domain-specific topics pertaining to security. Publications within the series are peer-reviewed monographs and edited works in the areas of:

- biological and chemical threat recognition and detection (e.g., biosensors, aerosols, forensics)
- crisis and disaster management
- terrorism
- cyber security and secure information systems (e.g., encryption, optical and photonic systems)
- traditional and non-traditional security
- energy, food and resource security
- economic security and securitization (including associated infrastructures)
- transnational crime
- human security and health security
- social, political and psychological aspects of security
- recognition and identification (e.g., optical imaging, biometrics, authentication and verification)
- smart surveillance systems
- applications of theoretical frameworks and methodologies (e.g., grounded theory, complexity, network sciences, modelling and simulation)

Together, the high-quality contributions to this series provide a cross-disciplinary overview of forefront research endeavours aiming to make the world a safer place.

The editors encourage prospective authors to correspond with them in advance of submitting a manuscript. Submission of manuscripts should be made to the Editor-in-Chief or one of the Editors.

Peter D. E. Biggins · Deeph Chana
Editors

CBRNE: Challenges in the 21st Century

Springer

Editors
Peter D. E. Biggins
Institute of Security Science and Technology
Imperial College London
London, UK

Deeph Chana
Institute of Security Science and Technology
Imperial College London
London, UK

ISSN 1613-5113 ISSN 2363-9466 (electronic)
Advanced Sciences and Technologies for Security Applications
ISBN 978-3-031-17373-8 ISBN 978-3-031-17374-5 (eBook)
https://doi.org/10.1007/978-3-031-17374-5

This Springer imprint is published by the registered company Springer Nature Switzerland AG
The registered company address is: Gewerbestrasse 11, 6330 Cham, Switzerland

Preface

Context for This Book

At the time of preparing this book, the world was going through the profound event of a pandemic in the shape of the COVID-19 virus and the planet's population has become increasingly aware of the global and long-lasting impact that such events can have. In many aspects, the virus has revealed national and international shortcomings in assessing, communicating and mitigating risk, and furthermore, its inability to coordinate responses on local, regional, national and international scales. Importantly, for this book, COVID-19 has also sparked a debate as to the virus's origin, with information, misinformation and disinformation circulating around the theories of this biological event's natural vs human manufactured origin story. This and other memes have greatly amplified the impact of the virus beyond its physical health consequences, polarising communities, bringing about political upheaval and calling into question the validity of science and scientific reasoning in many circles. In a cyber security obsessed security narrative, it has been an important reminder to us that we live in a cyber-physical world in which threats to our security and resilience are now characterised by complex interactions and interplays between our material reality and our digital existence. In the modern world, physical threats are characterised by cyber enablers and cyber consequences whilst cyber threats are, similarly, enabled by physical consequences and physical enablers; a key example of this point being the hardware infrastructure that connects us over wires, fibre and radio waves. Furthermore, if we examine the trends of this infrastructure, we can readily see that the coming years and decades will see a continued increase in our global interconnectivity with both digital and material objects, including ourselves, moving more freely over our networks and systems.

Structure of This Book

We have already alluded to the fact that in the world of security thinking there is often a line of demarcation drawn between the so-called 'physical' security problems and the domain of 'cyber' security. In this traditional taxonomy of security problems, physical security threats are typically, in turn, further categorised as being Chemical (C), Biological (B), Radiological (R), Nuclear (N), Explosive (E), collectively referred to by the initialism, and the CBRNE acronym is widely used throughout this book. Although N is mentioned, it is not covered in detail. In the early chapters of this book, we have invited contributions from authors with experience of working in each of these categories with a remit to provide historical context of the subject matter and bring the reader to a contemporary understanding. Each one of these chapters may be read in isolation if a primer is needed in any one of the CBRE elements, but in combination, it is hoped that the reader will be able to gain a good foundational understanding of the overall CBRNE security landscape and the linkages between each of the categories. On this last point, the reader should become aware that the taxonomy of physical threats is an imperfect one, and whilst useful for providing a framework for managing and working on such threats, there are no rigorous delineations that we feel are convincing from a scientific standpoint. Radiological materials are obviously chemicals as are explosives and it is not entirely clear whether to classify viruses as chemicals or biological entities—although they are typically treated in the latter category. We point out the imperfection of this taxonomy not to criticise its use, but rather to warn against an over obsession with its exactness, which is worth criticising.

Following a discussion of CBRNE, we then spend several chapters discussing the topic of security risk with the aim of illustrating just how challenging it is to make quantitative assessments of risk and relative threat severity when we consider physical threats in our modern world. It is seductive to think that the calculation of such risks is relatively straightforward as, naively, all that seems to be needed is the quantification of factors such as an event's likelihood and its impact. The discussions that are exposed by our authors, however, show that deriving an understanding of each is far from trivial and uncertainties abound. As an illustration of this, if we simply recognise that human behaviour is extremely complex and difficult to formulate, we realise that risk quantification requires a partnership of subjective and objective considerations that makes it far from an exact discipline. Furthermore, the increasing complexity of our cyber-physical world seems to exacerbate the situation, introducing novel factors and emergent behaviours for consideration that may not previously have been considered. This issue of determining risks accurately leads naturally to difficulties in assessing how we might mitigate or mollify CBRNE risks. Once again it is human factors that, arguably, contribute most heavily to the full weight of this task, which lies at the core of transforming situations of insecurity and fragility into those of security and resilience. Chapter "Human Factors and Societal Aspects in Future CBRNE Incidents", therefore, expounds on the topic of utilising human factors in crisis response, focusing on how CBRNE impact may be addressed from this perspective.

In the final third of this book, we hope to illustrate and explain how physical threats are now embedded in modern socio-technical settings where humans and machines interact with each other through data, algorithms, and physical actuation. The ubiquity of cyber-physical technologies such as driverless cars, autonomous delivery systems, smart homes and smart-healthcare devices, amongst other innovations, should compel those who consider problems of cyber security and physical security to be unbridgeable silos to think again. The taxonomy of physical vs cyber security needs to be recognised as being an imperfect one, with more and more of the interesting and societally meaningful problems lying in their intersection. At the time of writing, it seems that communities typically focused on 'cyber' have recognised the need to step away from such dogmatic adherence to these labels, but the same cannot necessarily be said of agencies, organisations and practitioners who consider themselves experts in CBRNE related matters. Another traditional boundary between security communities that also proves to be counterproductive—through examples such as the COVID-19 pandemic—is that between those who consider questions of psychology and human behaviour and those who consider more tractable analytic problems in technology. Our digital world now enables the mass propagation of disinformation and the generation and global distribution of misinformation, both with the potential to cause emotional, mental and, most importantly for our discussion here, physical damage and impact. In one form or another, we can suppose that disinformation campaigns are part of the military thinking of nearly every nation state and many non-state organisations too. This motivates the urgent need for research communities and practitioners who are well versed in human behavioural research in combination with in-depth technical knowledge, such as the generation of deep-fakes using machine learning.

At the time of writing the concept of digitally created virtual realities, termed metaverses, is perhaps the most disruptive socio-technical trend that is set to shape human existence in the decades to come. Taking early ideas of virtual reality to another level, metaverses promise the ability to decentralise organisations, transact real contracts and agreements and exchange value and assets through cryptocurrencies and non-fungible tokens. The blurring of boundaries between material and digital realities that is at the heart of this technology evolution only serves to amplify the issues related to the erosion of current security taxonomies as the lines of demarcation between categorical boundaries become increasingly meaningless, acting as a prohibitive rather than an enabler to examining and solving security problems. In short, this book aims to outline the urgent need for CBRNE threats to be considered within the context of the emerging and disruptive technology trends that are shaping our societies.

London, UK

Peter D. E. Biggins
Deeph Chana

Acknowledgements The editors would like to thank the chapter authors for their contributions and for bringing the concept of this book to fruition. The contributing authors and editors also thank their colleagues and families for their support in this endeavour. In addition, special thanks go to Jack Cooper for obtaining permissions and proofreading and to Caroline Biggins and Max Swinscow-Hall for additional proofreading.

Contents

Contributors

Biggins P. D. E. Institute for Security Science and Technology, Imperial College, London, UK

Blatny J. M. Norwegian Defence Research Establishment (Forsvarets Forskningsinstitutt (FFI)), Kjeller, Norway

Brattekås K. Norwegian Defence Research Establishment (Forsvarets Forskningsinstitutt (FFI)), Kjeller, Norway

Chana D. S. Institute for Security Science and Technology, Imperial College, London, UK

Endregard M. Norwegian Defence Research Establishment (Forsvarets Forskningsinstitutt (FFI)), Kjeller, Norway

Goldbeck N. Department of Civil and Environmental Engineering, Imperial College, London, UK

Hassard J. F. Department of Physics, Imperial College, London, UK

Lavelle M. Intelligent Ontologies Ltd. & Institute for Security Science and Technology, Imperial College, London, UK

Nieuwenhuizen M. S. CBRN Protection Department, TNO, Rijswijk, The Netherlands

Nystuen K. O. Norwegian Defence Research Establishment (Forsvarets Forskningsinstitutt (FFI)), Kjeller, Norway

Ochieng W. Y. Department of Civil and Environmental Engineering, Imperial College, London, UK

Proud W. G. Blackett Laboratory, Institute of Shock Physics, Imperial College, London, UK

Sellevåg S-R. Norwegian Defence Research Establishment (Forsvarets Forskningsinstitutt (FFI)), Kjeller, Norway

Wilson M. A. Institute for Security Science and Technology, Imperial College, London, UK

The Changing Landscape

P. D. E. Biggins, J. F. Hassard, and S-R. Sellevåg

1 A Brief History of Crises and Armed Conflicts

The 20th Century demonstrated the human capacity for causing considerable loss of life as highlighted by the two World Wars along with all the other regional conflicts that have occurred. World War 1 caused approximately over 20 million deaths with 9.7 million military deaths [1] occurring in battles in contrast to the vast majority of deaths in previous centuries' conflicts occurring through disease [2]. The subsequent and arguably consequent H1N1 pandemic, commonly known as Spanish Flu, caused an estimated 50 million deaths. By contrast, the Second World War caused up to 85 million deaths with perhaps 25 million suffered by the militaries of the many countries involved and the Soviet Union and China being by far the countries most afflicted. Towards the end of that century there was a brief period of optimism that occurred when the Berlin Wall, which was a symbol at the time of the 'bipolar world', fell in November 1989 bringing the end of the Cold War between the USA and the USSR. However, the reality of the Balkan conflict provided a check on that optimism. Since 2010, as shown in (Fig. 1) there has been a steep increase in the number of intrastate and internationalised intrastate armed conflicts in the world [3]. Many would now regard the present multipolar nuclear-armed and chemical/biological-capable world is inherently less stable than that of just 40 years ago. The Russian

P. D. E. Biggins (✉)
Institute for Security Science and Technology, Imperial College, London SW7 2AZ, UK
e-mail: p.biggins@imperial.ac.uk

J. F. Hassard
Department of Physics, Imperial College, London, SW7 2AZ, UK
e-mail: j.hassard@imperial.ac.uk

S-R. Sellevåg
Norwegian Defence Research Establishment (Forsvarets Forskningsinstitutt (FFI)), P.O. Box 25, Kjeller, Norway
e-mail: stig-rune.sellevag@ffi.no

P.D.E. Biggins and D. Chana (eds.), *CBRNE: Challenges in the 21st Century*, Advanced Sciences and Technologies for Security Applications,
https://doi.org/10.1007/978-3-031-17374-5_1

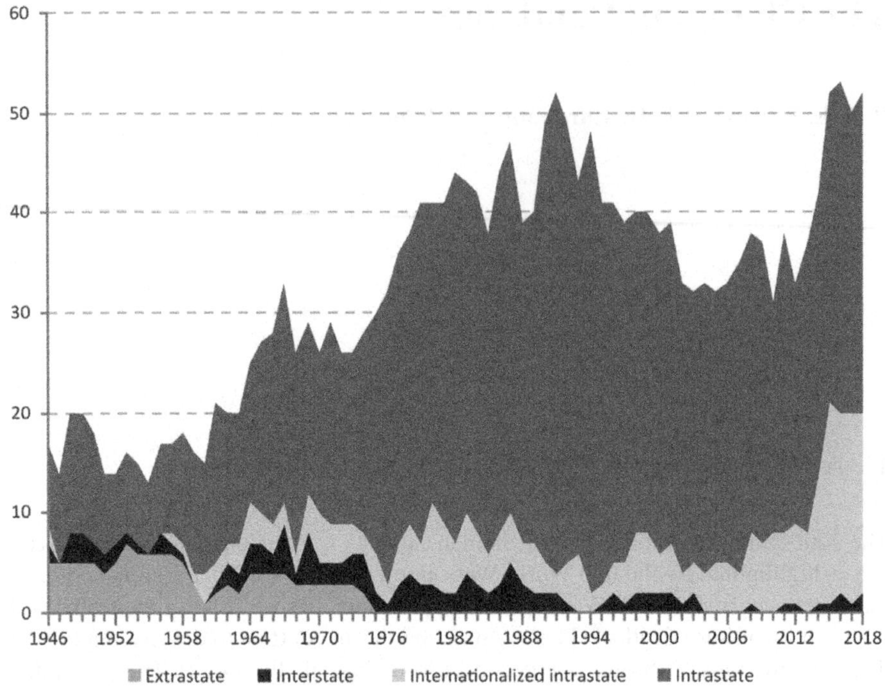

Fig. 1 Armed conflicts in the world by type and year (1946–2018). *Source* Pettersson et al. [3] courtesy of Sage Publishing

invasion of Ukraine on the 24th of February 2022 graphically illustrates this point. Increasing pressures—primarily environmental but also population, societal division and creed-related—can only increase those instabilities.

The graphic horror of 11th September 2001, known as 9/11, heralded in the 21st Century with a stark reminder that there is considerable disunity in the world. The attacks were not only a shock to the USA, but to all Western countries, resulting in permanent security measures and the consequent wars in Afghanistan and Iraq. Regardless of whether 9/11 can be characterised as a strategic surprise event (or a "Black Swan" type of event) or as a failure to respond to emerging threats, it highlights the importance of being able to manage surprises and identify and recognise weak threat signals (often correlated with societal pressures) in order to safeguard our security. The first decade of the new millennium was not only shaped by the so—called "War on Terror" [4] but also by the 2008 global financial crisis that started in the USA in late 2007. The subprime mortgage crisis turned into a global economic recession due to the highly interconnected nature of the global financial system. Although the individual risks were viewed as small, the systemic risk to the financial system was significant. Furthermore, our inadequate understanding of the risks, particularly their correlated nature, and the effects are still affecting many countries today which are being compounded by the 2020 COVID-19 global pandemic.

The Arab Spring that began in Tunisia in December 2010 and Libya in February 2011 initiated uprisings against authoritarian rule in several countries across the Middle East and North Africa. Also, unrest started in Syria and escalated to an ongoing civil war at the time of writing in 2021; the second deadliest war in the twenty–first century with numerous reported incidents of chemical weapons use. The Syrian Civil War spread violence into Iraq and sparked the rise of the Islamic State and other violent extremist groups.

2 A Changing World Order

Globalisation, urbanisation, climate change and technological development continue to change our societies. Today we live in a world where the exchange of people, goods, money, information and ideas bring us new opportunities, services and benefits for humanity. Equally, some threats and exploitations have increased exponentially with this growing societal complexity.

Throughout the world, urbanisation is growing, and the density of populations have been rising since the 1800s. It is predicted that more than two thirds of the global population will live in urban areas by the year 2050, 64% of the developing world and over 86% of the developed world [5]. In (Fig. 2), the projected growth of the global population indicates that approximately 6.7 billion people will be living in urbanised areas as opposed to 3.1 billion in rural areas. How future urban growth in various parts of the world will occur in terms of spatial development and what its implications will be for human well-being and the environment in the light of the recent COVID-19 pandemic is hard to predict.

In fact, the impact of climate change will cause populations to migrate from arid or flood areas creating challenges at the borders of neighbouring countries. In the last 60 years, atmospheric CO_2 has risen from some 250 ppm to 417 ppm. The consensus amongst expert climate scientists is that this is entirely due to anthropogenic activities, primarily the burning of coal, oil, and gas, all from fossil sources. These sources were entirely as a result of solar activity over millions of years laying down complex hydrocarbons which stored in a small fraction of the energy from the sunlight from all those millennia ago. That elevated level of CO_2 will inevitably cause rises in temperature on a global basis, which at the poles could exceed 8 °C, and even at the Equator at least 1.5 °C. Ice melting at the poles, and the expansion of the seas, will cause significant sea level rise, disrupting hundreds of millions of people. Temperature rises at the Equator will make large swathes of the globe uninhabitable. For example, in the Middle East North Africa (MENA) region, it is likely that over 350 million people will find their homelands uninhabitable within the next 50 years. Bangladesh is arguably the most vulnerable country on earth; over 70 million people live within 2 m of sea level. Their forced migration could result in the largest natural calamity in history, compounded by related migrations from other countries nearby.

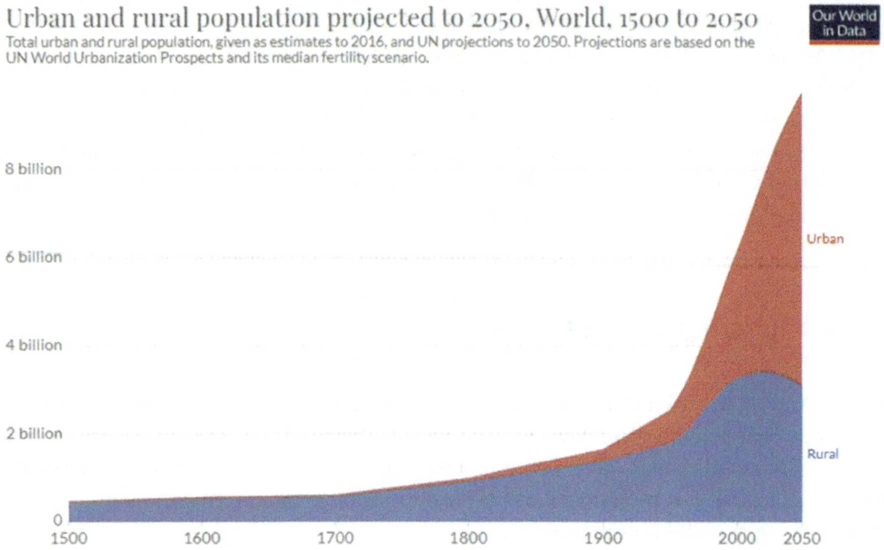

Fig. 2 Urban and Rural population projected to 2050 (*credit* Our World in Data Creative Commons BY license)

There is a potential shift in global transition of power in Asia due to the emerging economies, potentially rivalling that of the West. From a Western perspective, the current international system with its institutions, mechanisms and 'norms' may find it challenging to accommodate this potential shift in economic power. If this occurs, it may become more difficult to resolve disputes, which could lead to an increase in trade wars with the potential risk of conflicts, especially if the governments take a more authoritarian stance.

The transition of power does not only occur between states; there is also a diffusion of power within the states and to non-state actors. With the trend of urbanisation, for example, the economic, demographic and political power of large cities may grow substantially relative to the state, which could further increase the inequality between the rich and the poor. In addition, a few multinational corporations may become so powerful in the future due to their control of vital services that they provide to the state, that there is a risk that they will use their influence to demand concessions from the state. A question in point is who controls the communication networks? The technological high frontier can also be the 'ersatz' battleground between incumbents and those countries which aspire to challenge those incumbents often in the name of commerce, (for example, in 5G) but usually with a security-related flavouring.

3 Coping with Complexity

Life as we have come to know it in the modern world is becoming increasingly dependent on critical infrastructures for maintaining vital functions in our society. Already the advances in digital technology are moving at such a rapid rate that it is transforming our societies into so–called "smart societies". The concept of a smart society revolves around the implementation of a digital backbone of technologies within the infrastructure that improves economic efficiency, communications, the response and effectiveness of organisations, health care provisions with the overall benefit to the society's well-being. The COVID-19 pandemic has revealed, for instance, the heavy reliance and need for a robust digital communications infrastructure. This will have considerable influence on the way society and the commercial interests will operate in the future. It is obvious that COVID-19 is by no means the last time a pandemic will sweep the entire world, as history has shown us. With the combination of global travel aiding rapid transmission of disease and climate change affecting natural habitats and zoonotic disease reservoirs along with human vector migration, there is a need to fully understand the impact of these complex interactions.

The ability to have sensors embedded into the fabric of urban landscape provides considerable benefits but has the potential to create vulnerabilities and can complement and de-bias human perception. However, this type of development would allow not only bio-surveillance applications to be able to test, track and isolate during epidemic and pandemic situations, but also better monitoring of air quality and early warning of natural hazards like flooding.

Whilst developments in digital technologies are helping boost the efficiency of society, they are also creating levels of complexity. The use of the mobile phone, the internet and other electronic communication services is becoming ubiquitous throughout the world enabling a vast majority of people to be connected. As electronic communication services are being used for increased functions in society, this is also creating situations where fewer people appreciate just how interwoven and comprehensive everything is becoming resulting in a growing lack of knowledge of the wider implications amongst users and policy makers. An example is information whether true or erroneous can be disseminated via social media platforms quickly creating potentially conflicting and confusing pictures of the ground truth.

Lessons from recent disasters such as hurricanes, large-scale power outages, terrorist attacks and the COVID-19 pandemic have shown that infrastructure disruptions can directly or indirectly affect other infrastructures through a complicated web of interdependencies across different industry and business sectors. Thus, the effect of significant infrastructure disruptions can impact large geographical regions and send ripples throughout the national and global economy. The situation is exacerbated not only by the impacts of climate change, but also by digital technologies, increased electrification and the proliferation of the Internet of Things that continue to add complexity to our Critical National Infrastructures (CNIs): no one single entity is in control of the system. This is also the case at the government level where the responsibility is spread out across several departments and ministries. CNIs are open sociotechnical systems that are influenced by inward and outward flow of goods, services and capital, as well as undergoing constant interaction and exchange with their economic, social and natural environments. As such, CNIs are characterised by complexity, emergent behaviour, limited functional decomposition, distributed control, indeterminateness and nonlinearity. This makes it difficult for CNI operators and national authorities to carry out risk assessments or to gain situational awareness during CNI disruptions. Coping with complexity is therefore of importance when informing policy makers and decision makers on national security issues related to CNIs. Improving the resilience of our infrastructure systems therefore becomes a matter of urgency along with provenance of data and the use of trusted systems.

Given the rapid development and pressure for tools to assess such risks, and the innovative technologies which can result, we often see a lag phase between technology implementation and the establishment of necessary safety and security strategies, policies and regulations. Controlling this problem by regulating disruptive technologies has proven difficult due to the pace of change in society and technology, a lack of legal precedence, and the international nature of deployment and operations. In addition, security aspects related to the digital transformation are not limited to the cyber-physical domain, but also to the social, political and economic domains to which the digital technologies interact. While we are mindful of the potential catastrophic impact of innovative technologies which can be used as weapons, we are also aware that disruptive technologies need not be a product of the Internet age. Biological and Chemical warfare agents have been used for thousands of years, often with great effect. It is not always clear that controls to problems associated with these 'old technologies' are any more advanced than those defending the cyber-physical domain. The 2001 anthrax attack in the USA, which used the existing infrastructure for delivery, led to "one of the largest and most complex [investigations] in the history of law enforcement" [6], but took nine years to conclude. The economic fallout of the clean-up operations, the criminal investigations and the Biowatch [7] programme that was initiated afterwards is in the order of billions of dollars.

The responses of nations and their ability to absorb shock of such events will depend on their level of resilience and this will vary significantly from nation to nation. Some nations will be in a more fragile state than others and therefore less able to respond. The concept of anti-fragility has been proposed such that by understanding the shocks and stressors impacting on a system, anti-fragility within systems can be developed by design, by building systems which can adapt and improve from the environmental shocks and stressors encountered [8].

There is a growing perception that society does not understand the impact of complexity in relations between risk and society, a fact made apparent by responses to many crises in the past. Furthermore, failures in dealing with the second- and higher-order effects can be far more devastating than the direct effects of the disaster. This situation is exacerbated by the fact that disasters in complex systems like nature or society often are better represented by a power law than a normal distribution. The reason for this is that in systems following a power law, extreme events will have higher probabilities than in systems following a normal distribution. Thus, we need to be more concerned about hazards following a power law than those represented by

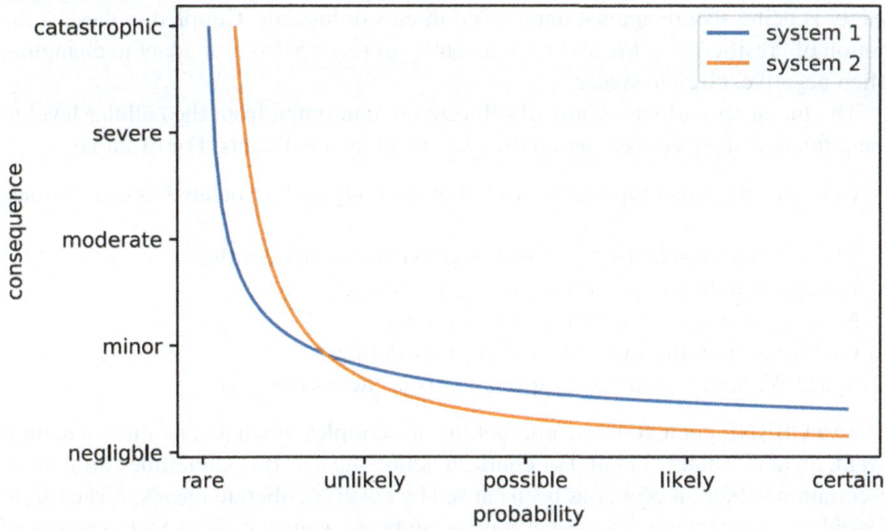

Fig. 3 Examples of power law functions for disasters in systems with different level of complexity and resilience. System 2 will be more resilient towards high probability, low-impact events than system 1, but more vulnerable to low-probability, high-impact events due to a higher level of complexity. (courtesy S-R. Sellevåg)

normal distributions. Although many steps have been taken to improve the resilience of our societies, one might think that not enough attention has been given to low-probability events with high consequences. When the level of complexity increases, the consequences of extreme, low-probability events will likely become more severe because of cascading effects as illustrated in Fig. 3.

4 The Multifaceted Threat Space

4.1 Threat Space Levels

From a security perspective "defence in depth" is a term that is often used where security of an asset is achieved by multi-layering of protective measures. An asset can be people in terms of employees, customers, contractors, friends and families or property both tangible and intangible that have an assigned value; examples of intangible assets include reputation, or information such as databases, software code, profiling data from social media platforms and critical company records. The protection of that asset is a basic human need. A threat to an asset can be something that can exploit a vulnerability deliberately or accidentally. The distinction between security and safety of an asset is that security is the protection against deliberate threats whilst safety is being secure against unintended threats or hazards. Coupled to these is the notion of "resilience"—the ability of an entity to recover from or adapt to changing, often negative, circumstances.

The threat space that we are all subjected to can range from the cellular level to the influence of space weather on the planet and its inhabitants. For example:

Cellular—the subcomponents of the human body and associated vulnerabilities, etc.
The Individual—identity, personal possessions, vulnerabilities.
Groups of Individuals—influence, vulnerabilities.
Nations—political instability, internal conflict.
Global Community—global conflict or pandemics.
Space Weather—solar flares, space debris or meteorites.

From a CBRNE context it is worth looking at examples of what constitutes a natural event, an accidental event and of course a deliberate act. It is sometimes difficult to ascertain whether an event has been caused by covert deliberate means. It is entirely feasible for a lone actor, groups harbouring umbrage against society, State-sponsored groups or States to create an event that on the face of it has been caused by natural or accidental actions. Furthermore, natural hazards may trigger and exacerbate technological disasters, so—called "Natech" disasters, like the Fukushima Daiichi nuclear disaster.

In a similar, rapidly evolving example of system complexity and criticality influencing a set of threats we must also consider the relationships between cyber-attacks—presently underway across the world—and the security, safety and resilience of existing societal infrastructures, such as chemical plants, water supplies and indeed nuclear installations, all of which could be configured to appear accidental, or overt.

4.2 Natural, Accidental and Deliberate (NAD) Events

4.2.1 Natural Events

We are being constantly subjected to natural phenomena due to where we live or events such as earthquakes, lightning strikes, hurricanes, tsunamis, flooding, solar storms and diseases, although it is very much dependent on location that will dictate how frequently these events are experienced.

From a chemical, biological and radiological perspective, we are subjected to natural events. These can occur in vastly varying timescales, which can disguise their intensity, and even their presence. Some preindustrial climate changes, all naturally occurring, led to the slow, but inevitable end of societies. For example, natural variations in rainfall arguably led to the end of several Mayan societies [9].

Other chemical processes are much more rapid. For instance, the outgassing of 1.6 million tons of carbon dioxide from the bed of Lake Nyos in Cameroon in 1986 resulted in the death of 1,746 people and over 3,500 livestock due to suffocation.

Similarly, our exposure to radiation occurs due to the earth being bombarded by a stream of cosmic radiation and the doses we receive will depend on location, elevation, atmospheric conditions and the Earth's magnetosphere. Radioactive materials such as uranium, thorium and radium exist naturally in soil and rock; for example Cornwall in the UK is designated as a radon affected area.

Turning to natural biological hazards, some diseases can extend globally. For example, the world has experienced pandemics, notably the Great Plague (also known as the Black Death) caused by the plague bacterium *Yersinia pestis* during 1347–1351 resulting in the estimated death of between 75 and 200 million and a significant impact on European history due to the economic effects and labour shortages. Similarly, economic effects were observed during the Influenza Pandemic (also known as Spanish flu), H1N1, during 1918–1920 with an estimate of between 50 and 100 million deaths. The Ebola outbreak of 2014–2016 highlighted the transmission of viruses from bats to humans and the current global pandemic COVID-19 crisis has yet to establish the origin although bats have been suggested as a potential source.

When natural events such as earthquakes, hurricanes, tsunamis, occur they can also trigger significant technological events, for example, damaging oil and gas pipelines, waste sites, power generation plants, or cause power surges, communications network blackouts, electronic systems etc. One of the most recent events that demonstrate this is the Fukushima Daiichi nuclear disaster caused by the Tōhoku earthquake and tsunami in 2011, which resulted in three nuclear meltdowns, three hydrogen explosions and the widespread release of radioactive contamination (*caveat:* the Fukushima incident is also regarded as an accident) [10].

4.2.2 Accidental Events

What we mean by an accident is usually a happening of an event by chance, unintentionally or unexpectedly. Over the years, there have been many accidental chemical incidents that have stemmed from failure of understanding and technology. The consequences have resulted in the huge loss of life as witnessed at Bhopal, India in 1984 when the Union Carbide pesticide plant accidentally released more than 40 tons of methyl isocyanate and other gases causing at least 3,800 deaths and affecting over 550,000 people [11]. Another example is the poisoning of nearly nine million people in the state of Michigan, USA in 1973 with polybrominated biphenols (PBB) due to a chemical mix-up which resulted in PBB's being put into the food chain [12], the effects of which are still detectable today.

Similarly, there have been several biological accidents that have happened. In 1978, the accidental release of the Variola virus (commonly known as smallpox) from a laboratory in Birmingham (UK) [13], resulted in the death of one person who worked nearby. The accidental release of Anthrax spores at Sverdlovsk, Russia, in 1979, which resulted in 105 deaths was due to a maintenance failure in the filtration system [14]; it was also a serious violation of the Biological Weapons Convention (BWC). In both cases poor maintenance of containment facilities led to loss of life.

In 2001, a farmer fed pigs with untreated waste that led to the UK Foot and Mouth disease disaster resulting in over £6.3 billion economic loss due to the impact on agriculture and tourism [15]. In 2005 and 2008, three outbreaks of Legionnaires' disease in the Fredrikstad/Sarpsborg community, Norway resulted in 61 cases of the disease, of which 12 were fatalities. The outbreaks were caused by the release of Legionella pneumophila to air and to the river Glomma from aeration ponds of a biological wastewater treatment plant at Borregaard Ind. Ltd., Sarpsborg, Norway [16, 17].

In 1986, the Chernobyl nuclear power plant in Prypiat, Ukraine suffered a nuclear meltdown in the graphite moderated reactor number 4 due to a safety test going wrong. A steam explosion occurred, and fire killed up to 50 people with estimates of 4,000 fatalities and of several 100,000 additional cancer deaths over time. Fallout could be detected as far away as Canada. The Chernobyl Exclusion Zone, covering portions of Belarus and Ukraine surrounding Prypiat, remains contaminated and mostly uninhabited. This nuclear accident resulted from the concurrence of several major factors, including the activities of its operators, the physical characteristics of the reactor and the design of its control elements, an underlying deficient safety culture and the fact that the reactor was brought to a state not specified by procedures which led to its unstable behaviour [18].

4.2.3 Deliberate Events

Deliberate acts of the use of chemical, biological and radiological agents are well documented. The first World War (WW1) demonstrated the industrial scale of the use of chemical weapons by Nation States and the development of doctrine and concepts of use by their armed forces. There has been sporadic use of chemical weapons in regional conflicts and civil wars since WW1, especially in the Middle East and most recently in the Syrian Civil War.

Chemical, biological, and radiological agents have also been used in deliberate assassinations and assassination attempts. In 1978, the Bulgarian dissident Georgi Markov was assassinated in London from a small ricin filled pellet fired from an umbrella into the back of his right thigh. A similar attack in Paris, a couple of weeks earlier, was carried out but fortunately the victim survived. In 2006, the former Russian FSB and KGB officer Alexander Litvinenko was assassinated by polonium-210 which was added to his cup of tea. The most recent incident occurred in Salisbury, UK in 2018 when an attempted assassination using a novel nerve agent known as Novichok was carried out against the exiled Russian GRU officer Sergei Skripal and his daughter. The assassins discarded the perfume bottle containing the nerve agent which subsequently led to the death of a British national a few months later in Amesbury, UK. In September 2020, the German Chancellor Angela Merkel stated that an attempt was made to poison the Russian opposition activist Alexei Navalny with a nerve agent.

Chemical and biological agents have also been used for terrorism. Some historic examples include the 1984 food poisoning attack carried out the Rajneesh Movement in The Dalles, Oregon, USA, the use of sarin by the Japanese cult Aum Shinryko in 1994 and 1995, and the Anthrax letters (also known as Amerithrax) in the USA in 2001. It is worth noting that in the case of the Rajneesh Movement attack, it only became known after a year when one of the members admitted they had carried it out [19]. Similarly, in the case of the Anthrax letters a considerable amount of ambiguity is associated with attribution [20].

4.3 Nuclear Proliferation

Since the atomic bombings of Hiroshima and Nagasaki in 1945 and the following nuclear arms race between USA and USSR during the Cold War, the use of nuclear weapons has posed an existential threat to humankind. The trust that was established during the 1990s between the two major nuclear powers, USA and Russia, and which led to significant reductions in the global stocks of nuclear weapons, has eroded gradually and resulted in the breakdown of several nuclear arms control regimes. This situation, together with the realisation of North Korea as a nuclear weapons state, the continued tensions between Pakistan and India, the attempts by Iran to develop sensitive nuclear technology outside of IAEA's verification regime, and the growing interest in nuclear power in the Middle East and North Africa (MENA) region, may point to a second era of nuclear arms racing. The proliferation of reliable information on the development, design and production of nuclear weapons as well as renewed interest in so−called "battlefield" nuclear weapons by established nuclear weapons states, only add to the threat posed by nuclear weapons. Actions are therefore required to prevent further nuclear proliferation, in particular ensuring control over and reducing the availability of weapons-usable fissile material.

4.4 Pandemics

The COVID-19 crisis of 2020 is a stark reminder that diseases know no boundaries and that no one is immune from naturally occurring diseases. It is only through vaccinations, antibiotics, antivirals, good medical practice, and herd immunity that we can overcome emerging, persistent, and neglected diseases. The COVID-19 pandemic has illuminated vulnerabilities not only in the health of the global population but the disparities within communities and the second order, and higher, global societal and economic impacts.

Whether or not governments took the right course of action at the time will be discussed for some time, however, a common strand that will be recognised is the interdependency of our global connectivity through our global socio-economic interactions. Whilst the disease itself has a clearly defined first order effect Nation States find themselves having to deal with the knock-on effects it has on their way of life. Although the Foot and Mouth disease that occurred in the UK in 2001 was not a pandemic, it illustrated the first and second order effects of a disease outbreak in terms of the devastating impact on the livestock industry (over 6 million cows and sheep were culled) along with the mental wellbeing of the farmers, vets and those dealing with it. The loss was not just in agriculture estimated to be £3.1 billion, but the knock-on effect was to close the UK countryside which led to the losses of between £2.7-£3.2 billion for the UK tourism industry [15]. The cost of the COVID-19 global pandemic will be in the order of trillions of dollars [21].

Until definitive evidence is established over the source of COVID-19, debate will continue as to whether COVID-19 was a natural event, an accidental crossover from animal reservoirs, an accidental release from a research laboratory, or an abhorrent act of deliberate dissemination. However, the message is clear: our global societies are vulnerable to a wide range of natural, accidental and malevolent events. Although the previous Middle East Respiratory Syndrome (MERS), Severe Acute Respiratory Syndrome (SARS) and Ebola outbreaks have provided ample learning opportunities and warnings of our vulnerability to a global pandemic, it took the reality of the COVID-19 pandemic to initiate mitigation measures. A clear message has been sent: develop appropriate global response and mitigation strategies.

The responses of nations and their ability to absorb shock of such events will depend on their level of resilience and this will vary significantly from nation to nation. Some nations will be in a more fragile state than others and therefore less able to respond. The concept of anti-fragility has been proposed such that by understanding the shocks and stressors impacting on a system, anti-fragility within systems can be developed by design, by building systems which can adapt and improve from the environmental shocks and stressors encountered [8].

5 The Grey Zone

The rate of change of technology, the use of social media in innovative ways and political changes are destabilising the 'norms' of what was once considered daily life. A challenge facing societies is establishing the ground truth of events and the importance of timely, accurate and confident reporting enabling informed decisions to be made. Often, the political decision makers are reacting to events that are happening at pace and therefore their advisors need to have access to high-quality trustworthy information.

We are starting to see combinations of different actions arising from different quarters, such that we are becoming increasingly exposed to unorthodox and asymmetrical actions. This highlights that defence and security issues increasingly overlap and the once apparent distinction between the two are becoming less obvious. It is this blurring of the boundaries between security and defence that creates uncertainty.

This is captured well by General Sir David Richards when describing his future vision of wars [22]:

> State on state warfare is happening and will continue to happen, but some are failing to see how. These wars are not being fought by conventional invasion of uniformed troops, ready to be repulsed by heavy armour or ships, but through a combination of economic, cyber and proxy actions.' ....' the present shift is one that includes our entire society and therefore impacts our whole security infrastructure.

Whilst this quote places emphasis on combinations of economic, cyber and proxy actions, consideration of chemical, biological and radiological agents should be added to that cocktail. Given that the health and economic impacts of accidents and natural events are significant, it is conceivable that covert attacks may be disguised as accidental or natural events. The possibility of experimental accidents, deliberate release and perhaps more importantly of natural events presents *a priori* hazards to societies and serious risk of disruption to their economies with the attendant impacts.

The blurring of the boundaries between security and defence—often characterised as the 'grey zone'—is also reflected upon in France's 2017 defence and national security strategic review [23]:

> Technology development, advanced military capabilities dissemination, and the emergence of new domains of confrontation [cyber and outer space] present new challenges to traditional defence and security systems. By expanding the range of options available to all types of actors [state and non-state], they are facilitating the development of integral and grand strategies that mix military and non-military capabilities, as well as so—called "hybrid" postures based on a combination of ambiguous postures, direct and indirect pressures enabling flexible action from peacetime to open conflict. While the global development of military capabilities at both the upper and lower ends of the spectrum leads to harder armed conflicts, the risk of serious incidents or uncontrolled escalation is increasing.

Given the ambiguity of so-called grey zone activities or 'hybrid' threats, there is no consensus on the definition of such threats. Hybrid threats therefore constitute a 'wicked' problem. Still, it is acknowledged that grey zone activities constitute a significant threat to national security.

The European Centre of Excellence for Countering Hybrid Threats has characterised the phenomena as follows [24]:

- Coordinated and synchronised action, that deliberately targets democratic states and institutions systemic vulnerabilities, through a wide range of means.
- The activities exploit the thresholds of detection and attribution as well as the different interfaces (war-peace, internal—external, local-state, national—international, friend-enemy).
- The aim of the activity is to influence different forms of decision making at the local (regional), state, or institutional level to favour and/or gain the agent's strategic goals while undermining and/or hurting the target.

Thus, in order to understand grey zone activity as a phenomenon, it is necessary to understand the context in which it appears in, as well as its characteristics and drivers.

Some general characteristics can be identified: firstly, grey zone activities remain below the threshold that would justify a conventional military response. Thus, the aggressor seeks to avoid threatening the defender's vital or existential interest as well as avoiding attributable violations of international laws or norms. Covert, non-attributable attacks involving non-military means are therefore important characteristics for grey zone activities. At the same time, however, the aggressor exploits the risk of escalation to a direct military conflict. Another important characteristic which underpins the ambiguous, non-attributable features of grey zone activities, is that the aggressor tries to unfold its actions gradually over time rather than trying to achieve all objectives in one step since this would enhance the risk of a direct conflict. Lastly, the aggressor seeks to exploit specific vulnerabilities in the targeted country. Such vulnerabilities can be political polarisation, social unrest, economic stagnation, vulnerabilities in national defences, or vulnerabilities and loopholes in laws and regulations. Important drivers for the phenomenon often relates to the costs of weapons and conventional war, *i.e.*, the increasing political and economic costs of military warfare, the risk of escalation to nuclear and the rapid technological development of advanced (and therefore expensive) weapons such as long-range, high-precision missiles.

The blurring of boundaries is not only occurring between security and defence; we also see a blurring between fact and misinformation in the increasingly expanding and unregulated information space. Social media with their 'echo chambers' (where people only see and hear the views of those with similar opinions) may be exploited by malicious actors to polarise populations, erode trust in politicians and institutions, create uncertainty and fuel grievances. In addition, the spread of fake news through social media may spin conspiracy theories such that 5G was responsible for the spread of COVID-19. With increasing volumes of information and advances in digital technologies, it may become increasingly difficult for the general citizen to separate trustworthy from fake information.

6 Multifaceted Threats and Society's Response

The multifaceted nature of threats complicate both the causes of any given event, and the optimised response from those charged with defending society. It is obvious that threats and society's response and mitigation strategies are rarely, if ever, single-faceted. This requires society to build strategies which consider the invariably multi-faceted threats, their causes and effects as well as the multidimensional responses required. It behoves those who seek to help solve the multifaceted problems the world faces, which result in, for example, CBRNE activities, to consider the problem holistically and with as little bias as is possible.

A threat from, particularly, a well-resourced national or (sometimes) a subnational group will likely include thought on how to evade detection, at each stage, in what might be a complex sourcing and production cycle. It will be obvious that in a serious attack, for example one employing CBRNE, more than one agent could be deployed, making the response strategies not multiplicatively but exponentially more complex.

Imagine the scenario: first responders are dying in a subway attack, and sarin is identified [25]. The response may be to focus on that aspect to the neglect of other hazardous materials.

Indeed, the partial but deniable discovery of the plot might be exactly what the perpetrators and their bosses want. For example, it is not inconceivable that the use of the nerve agent Novichok, regarded as a "signature" attack mode, was chosen precisely to send a message.

7 The Landscape

This introduction has shown that we are facing a new era of threats and changes in global security that has an impact on both civilian society and national defence. Nation States are experiencing a plethora of threats ranging from ideological extremism, terrorism, migration issues, climate change effects, cybercrimes, diseases and economic instabilities and the growth of fake news to name but a few. From a national security perspective terrorism, espionage, cyber threats, and the proliferation of weapons of mass destruction are of significant concern, many of which impact on critical national infrastructure.

These threats are revealing societal vulnerabilities which are taxing academia, industry, and government thinkers. This has been exemplified by the increasing blurring between the defence and security sectors adding to a level of complexity and challenge to societies due to the spectrum of aggressors from lone wolf operatives at one end to that of nation states at the other. The global community is therefore facing a geo-political changing landscape that will create pressures on societies. From a societal perspective there is an expectation of a level of security as security is a basic human need whether it is in terms of protection of the individual, property, groups, organisations or the country against threats such as crime or attacks by foreign countries [26].

Understanding these changes, their consequences, and the societal vulnerabilities they carry, obliges a knowledge-based all of society approach where science and technology are crucial to providing insight and solutions.

References

1. https://pdf4pro.com/view/world-war-i-casualties-centre-robert-schuman-org-a6ff3.html. Accessed 21 September 2022
2. https://www.futuremedicine.com/doi/full/10.2217/fmb-2018-0323. Accessed 2 June 2021
3. Pettersson T, Högbladh S, Öberg M (2019) Organized violence, 1989–2018 and peace agreements. J Peace Res 56:589–603
4. https://2001-2009.state.gov/s/ct/rls/wh/6947.htm. Accessed 29 July 2020
5. The Economist. 27 October 2012. Accessed 20 March 2013
6. https://archives.fbi.gov/archives/news/stories/2008/august/amerithrax080608a

7. https://en.wikipedia.org/wiki/BioWatch. Accessed 6 Aug 2020
8. Taleb N (2013) Antifragile: things that gain from disorder. Penguin Books
9. Diamond J (2011) Collapse: how societies choose to fail or survive, 2nd edn. Penguin Books. ISBN 978-0-241-95868-1
10. Braun M (2011) The Fukushima Daiichi Incident, AREVA. https://hps.org/documents/areva_japan_accident_20110324.pdf. Accessed 7 June 7 2020
11. Broughton E (2005) The bhopal disaster and its aftermath: a review. Environ Health 4:6. https://doi.org/10.1186/1476-069X-4-6
12. Carter LJ (1976) Michigan's PBB incident: chemical mix-up leads to disaster. Science 192:240–243
13. Hawkes N (1979) Smallpox death in britain challenges presumption of laboratory safety. Science 203:855–856
14. Alibek K, Handelman S (2000) Biohazard: the chilling true story of the largest covert biological weapons program in the world—told from inside by the man who ran it. Arrow Books
15. Thompson D, Muriel P, Russell D, Osborne P, Bromley A, Rowland M, Creigh-Tyte S, Brown C (2002) Economic costs of the foot and mouth disease outbreak in the united kingdom in 2001. Rev Sci Tech 21:675–687
16. Olsen JS, Aarskaug T, Thrane I, Pourcel C, Ask E, Johansen G, Waagen V, Blatny JM (2010) Alternative routes for dissemination of Legionella pneumophila causing three outbreaks in Norway. Environ Sci Technol 44:8712–8717
17. Blatny JM, Reif BA, Skogan G, Andreassen O, Høiby EA, Ask E, Waagen V, Aanonsen D, Aaberge IS, Caugant DA (2008) Tracking airborne Legionella and Legionella pneumophila at a biological treatment plant. Environ Sci Technol 42:7360–7367
18. International Nuclear Safety Advisory Group (1992) The chernobyl accident, Updating of INSAG-1 (Safety Series No. 75-INSAG-7). International Atomic Energy Agency. https://www-pub.iaea.org/MTCD/publications/PDF/Pub913e_web.pdf
19. Miller J, Engelberg S, Broad W (2001) Germs: biological weapons and america's secret war. Simon and Schuster, p 382. ISBN 0 684 87158 0
20. National Research Council (2011) Review of the scientific approaches used during the FBI's investigation of the 2001 anthrax letters. Washington, DC: The National Academies Press. https://doi.org/10.17226/13098
21. Dobson AP et al (2021) Ecology and economics for pandemic prevention. Science 369(6502):379–381
22. Betz DJ, Stevens T (2011) Cyberspace and the state: toward a strategy for cyber-power. The International Institute for Strategic Studies, Routledge, p 80
23. Republic of France (2017) Defence and national security strategic review 2017. https://www.defense.gouv.fr/layout/set/popup/content/download/520198/8733095/version/2/file/DEFENCE+AND+NATIONAL+SECURITY+STRATEGIC+REVIEW+2017.pdf
24. European Centre of Excellence for Countering Hybrid Threats (n.d.) Hybrid threats. https://www.hybridcoe.fi/hybrid-threats/. Accessed 9 June 2020
25. Gupta RC (2015) Handbook of toxicology of chemical warfare agents. Academic Press, p 27. ISBN 9780128004944
26. Martin P (2019) The rules of security. Oxford University Press

Chemical Threats

J. F. Hassard and M. S. Nieuwenhuizen

1 Introduction

The threat of attacks using chemical, biological, radiological, or nuclear means (CBRN) describe some of the most extreme possible events in a scenario of conflict. This chapter focuses on chemical (C) threats, where chemical agents are intended to cause a toxic effect on human organisms and where our definition of toxicity excludes those related to radiation exposure and/or the actions of pathogens. We define chemical agents as compounds that may exist in any physical state (e.g., solid, liquid, aerosol, vapour, gas).

Historically, chemical threats have been divided into categories based on the physiological impact they cause or the target organ they attack, such as the skin, the eyes, the lungs, or central nervous system. Whilst understandable from a historical perspective, our modern understanding of biochemistry and physiology compels us to recognise such a scheme as being, generally, overly simplistic. Physiological effects are often extremely complex, with multiple body functions being affected, to varying degrees, by a single agent's action. Furthermore, if we consider the amount of a substance that might be absorbed as a factor, an even more complex picture of toxicity emerges. An extremely large set of chemicals is characterised by varying toxicity that is dependent on the level of dose; reminding us of the maxim widely attributed to the Swiss physician and alchemist, Paracelsus [1], "Sola dosis facit venenum" (only the dose makes the poison). Recognising that dose is a pivotal factor for consideration it becomes apparent that a range of other factors that describe a target organism's environment at a given moment in time, together with a chemical's intrinsic toxicity, combine in a non-trivial way to determine the impact per unit mass of a given agent.

J. F. Hassard
Department of Physics, Imperial College, London SW7 2AZ, UK

M. S. Nieuwenhuizen (✉)
CBRN Protection Department, TNO, Lange Kleiweg, 1372288 GJ Rijswijk, The Netherlands
e-mail: mnieuwenhuizen@kabelfoon.nl

P.D.E. Biggins and D. Chana (eds.), *CBRNE: Challenges in the 21st Century*, Advanced Sciences and Technologies for Security Applications,
https://doi.org/10.1007/978-3-031-17374-5_2

A vapour of a given substance, for example, will be affected by a myriad of time dependent environmental factors including turbulence, temperature and humidity, which will determine its chemical stability, concentration and dispersal. Chemical threat agents are, by character, extremely difficult to predict and control. A fact that has, unfortunately, been repeatedly established through the malevolent actions of humans on other humans through the ages.

2 Chemical Warfare

2.1 A Long History

In ancient times the concept of using toxic agents against military opponents as well as civilians was known and employed [2]. In many battles hundreds of years before the common era (BCE) all manner of toxic agents were used: from smokes containing arsenic, as an example, to the poisoning of water. Much later, the many suggestions, inventions, and concepts proposed during the American Civil War [3] were among the forerunners of chemicals that were subsequently developed to optimize the injurious effect of toxicity to humans as weapons.

The 1917 poem "Dulce et Decorum est" by Wilfred Owen [4] begins with the line "Gas! Gas! Quick boys" and provides a first-hand account of the use of chlorine gas on the front lines of World War I. Having experienced the full array of what we would now term 'conventional weapons,' Owen singles out the use of chlorine as being particularly inhumane due to the prolonged suffering it caused to its victims. According to Paxman and Harris [5] over half the artillery shells in World War I were filled with chemicals—typically mustard gas and phosgene. Hitler's regime continued the production of a stockpile in World War II and although refrained from using it on the battlefield, used chemical agents to devastating effect in concentration camps that killed about 6 million Jews.

On 22nd April 1915, Germany launched the first well documented large-scale chemical attack at Ypres on the Western Front, employing about 150 tons of chlorine gas. The German scientist Fritz Haber had proposed to the German High Command that an alternative weapon may help to create a breakthrough in the trench war. The agents chlorine (1774), hydrogen cyanide (1782), cyanogen chloride (1802), phosgene (1812), mustard gas (1822) and chloropicrin (1848) were subsequently used [6]. Although Haber was not a toxicologist, he influenced the science of chemical toxicology as well as the technology of chemical warfare. Haber conducted inhalation studies to optimize effects in chemical warfare.

He also developed Haber's Law or Rule which states:

> for a given poisonous gas, **tC=k,** where **C** is the concentration of the gas (mass per unit volume), **t** is the amount of time necessary to breathe the gas to produce a given toxic effect, and **k** is a constant, depending on both the gas and the effect.

Late in World War I other nations attempted to develop even more potent blistering agents, or vesicants, than mustard gas. Lewisite (dichloro-(2-chlorovinyl) arsine) was developed based on early research conducted in 1904 in the United States.

After World War I the chemical agents, phosgene and mustard gas, were considered amongst the most dangerous weapons of mass destruction developed. This changed with the discovery of nerve agents by Dr. Schrader of IG Farben in Germany when he accidentally isolated a compound called ethyl N, N-dimethylphosphoramidocyanidate, later known as tabun. Although such nerve agents were weaponized during World War II they were not employed. The Allies' response to the discovery of the German nerve agent programme after the war, however, was to develop their own programmes. This resulted in the US and the former Soviet Union creating some of the world's largest chemical weapons stockpiles, both of which have been significantly reduced through subsequent chemical warfare demilitarisation efforts.

Throughout the period that many historians refer to as the modern era, the use of chemicals on the battlefield has been a pursued tactic and topic of debate. Starting with the Strasbourg Agreement of 1675 that sought to limit the use of toxic agents there have been numerous treaties or elements of treaties aimed at controlling the use of chemicals in conflict. The concept of chemical warfare, as understood in the modern industrial context, emerged through this period with one of its first significant proponents being the British scientist and politician, Lyon Playfair who suggested the use of chemical payload artillery shells in the Crimean war in 1854. In the debate that followed this proposal Playfair had this to say to those who raised moral objections to any such idea:

> There was no sense in this objection. It is considered a legitimate mode of warfare to fill shells with molten metal which scatters among the enemy and produced the most frightful modes of death. Why a poisonous vapor which would kill men without suffering is to be considered illegitimate warfare is incomprehensible. War is destruction, and the more destructive it can be made with the least suffering the sooner will be ended that barbarous method of protecting national rights. No doubt in time chemistry will be used to lessen the suffering of combatants, and even of criminals condemned to death.

By 1899, however, Article 23 of the Convention of Laws and Customs of War on Land was established as an outcome of the First Hague Peace Conference, explicitly banning the use of 'poison or poisoned arms' and was ratified by many nations. A separate declaration banning asphyxiating gases in shells was signed by major European powers but rejected by the United States. Following World War I nations tried to ban the use of chemical agents. By 1925 the Geneva Protocol was signed pledging never to use gas in warfare again. The protocol did not involve verification regimes, nor did it try to destroy existing chemical weapons. As a result of this many nations continued to own and stockpile chemical weapons leading to numerous incidents of chemical weapon use over most of the next century. Eventually, in 1993, many countries—except for a small group—signed and ratified the Chemical Weapons Convention that banned all chemical weapons development, production, acquisition, stockpiling, transfer, or use and required the destruction of all existing stockpiles, all accompanied with a solid verification regime. After this most of the big stockpiles as well as so-called old and abandoned chemical weapons were destroyed and only a few 'rogue' nations possessed and used chemical warfare agents, including Iran, Iraq and Syria.

2.2 Chemical Warfare Agents (CWA) and Toxic Industrial Chemicals (TICs)

As discussed briefly above, chemical weapons are intended to enable attacks where the toxicity of a chemical compound or agent, either directly or indirectly, results in injury or impairment of human targets, often resulting in irreversible effects including death. Depending on the specific biochemistry that is exploited such agents can exact damage in a matter of seconds after exposure or over much longer time periods, sometimes with significant delayed impact. Chemical agents affect plant life and animals—including humans—by altering the chemical and biological properties of these organisms at the molecular level, attacking cells, pathways and biological systems, rather than damaging them through physical force, as is the case with explosives. Although we have already highlighted the limitations of doing so, it is still instructive to broadly classify such agents based on their primary toxic effects—noting that what is presented here is intended only for illustration:

- **Blister agents**: permanently damaging the skin, eyes or lungs e.g. mustard gas.
- **Oxygen-displacing substances:** directly impacting the function of the human body's circulation system e.g. chlorine cyanide and hydrocyanic acid.
- **Suffocating agents**: directly inhibiting or interfering with respiration functions and systems e.g. chlorine and phosgene.
- **Nerve agents**: causing disruption to the signal transmission between nerve cells and muscles e.g. sarin and VX.

Whilst the development and weaponizing of many of the chemical agent examples above have been as a result of large-scale military research activities, the increased use of chemicals in industrial processes has given rise to a new source of threat in recent decades. The rise in availability of Toxic Industrial Chemicals (TICs) has played a significant factor in lowering the barrier to entry for developing chemical weapons making such tactics feasible for smaller, less well-resourced groups and organisations such as terrorists. To date, chemical weapons have been employed by nation states, terrorist organisations, criminal groups and lone actors with horrific consequences and their use has been contemplated by every major power in the world. As the scale of the TICs production has grown globally, targeting the manufacturing sites of TICs has also emerged as a novel tactic for enacting chemical attacks.

2.3 International Structures Combatting CWA Use

Whilst historically there have been a number of agreements and conventions to prohibit the use of chemical weapons, a significant milestone in global control was achieved when Presidents George H.W. Bush and Mikhail Gorbachev signed the bilateral U.S.–Soviet Chemical Weapons Accord [7, 8] on June 1, 1990. This agreement was followed by the Chemical Weapons Convention (CWC) arms control treaty which entered into force on the 29th April 1997, which, implemented by the Organisation for the Prohibition of Chemical Weapons (OPCW) based in the Hague (The Netherlands), mandated the end of the development, production, stockpiling, transfer and use of chemical weapons, preventing their re-emergence and ensuring the elimination of existing stocks. The OPCW is a treaty-based organisation with responsibilities for the disarmament and non-proliferation and uses impartial mechanisms to verify compliance and to deal with non-compliance situations should they arise. As of 2022, based on OPCW data, 193 states have committed to the CWC, 98% of the global population live under the protection of the Convention and 98% of the chemical weapons stockpiles declared by possessor states have been verifiably destroyed. The OPCW won the 2013 Nobel Peace Prize.

The CWC distinguishes three classes of controlled chemicals making allowance for those that are used legitimately for industrial, medical or scientific purposes.

The three classes or "Schedules" are:

- ***Schedule 1***—Chemicals with no normal non-military use, such as sarin, ***lewisite*** and ***mustard gas***. Productions of more than 100 g and stockpiles of more than 1 tonne are disallowed by the OPCW.
- ***Schedule 2***—Chemicals with some small industrial uses such as ***thiodiglycol***, a solvent for inks, but also a precursor chemical of mustard gas.
- ***Schedule 3***—Chemicals with large-scale industrial uses. For example, chloropicrin is used as a fumigant. Plants producing more than 30 tonnes a year are regulated by the OPCW.

3 Chemical Terrorism

The attacks of September 9th 2001 in the USA are often seen as a turning point in sophistication of terrorist attacks by non-state actors, delivering large scale global impact. Chemical weapons and agents that come out of the classical chemical warfare domain may be considered as being suitable for enabling large scale terrorist impact. Where the chemical terrorist threat was previously seen as theoretical and, above all, small scale, the threat has become increasingly real, and the number of chemical attacks has increased in recent years. The increased dissemination of materials, technology, knowledge for making chemical weapons (agents and dissemination means), the lower threshold to obtain information played a role in this as well. In addition, it is assumed that the willingness among the possible perpetrators, ranging from small terrorist groups to state actors in asymmetric conflicts (i.e., Syria against its own population, almost certainly with help from outside actors) has increased. There is still some uncertainty about the probability and impact of chemical terrorism. Many governments state that it is a threat with a low probability and a high impact risk. The chemical variant of the various forms of CBRN terrorism is considered to be one of the most conceivable.

When chemical threat is not supposed to be large scale, there is an infinite variety regarding the "agent of choice," in fact referring to the old statement of Paracelsus [1] that in essence everything is toxic. Currently society has millions of chemicals with synthesis or production technology improving, including micro-scale synthesis, combinatorial synthesis and flow chemistry. Furthermore, many basic ingredients are simply available in the chemical industry or for sale in a complicated worldwide trade and logistic context. Finally, the knowledge for making chemical weapons is becoming widely available via the internet.

Recent incidents in which chemical weapons have actually been used are the use of sarin, chlorine and mustard gas in Syria and Iraq, among others. Improvised weapons such as so-called barrel bombs and roadside improvised explosive devices (IEDs) and chemical filled rockets have been developed. A special tactical deployment (in Syria) is that where a limited amount of chemicals (calcium hypochlorite, or hydrochloric acid with potassium permanganate) give the suggestion of a chemical weapon, which influences the operations, appealing among other things to fear. The associated knowledge and expertise is available on the internet or dark web or migrates with the spread of extremists. Interestingly, the use of chemical warfare agents is expressly forbidden in the Quran [9], and its associated accompanying texts, not least the Hadith from which most of Islamic Law is derived, reinforcing the notion that attempts to assign its use to Islamic forces is not only ill conceived, but counterproductive.

Some recent (murder) attacks illustrate the use of new generations of ultra-toxic non-volatile substances that enter the body via contact. The recent murder of Korean president's Kim Jung Un's half-brother with the nerve gas VX at Kuala Lumpur airport, as well as the incidents with a so called Novichok connection in Salisbury (United Kingdom), are examples of what could happen with highly toxic substances that were conceived in the military domain. Furthermore, there is in the chemical arsenal "cross-talk" from the pharmaceutical domain. The increasing availability of pharmaceutical-based agents such as the well-known fentanyl in the drugs world could easily belong to the terrorist arsenal.

Chemical terrorism can occur in various forms. From small-scale and improvised distribution in the open environment or in (critical) infrastructures of self-made or acquired resources to conventional attacks on warehouses or transport of resources that have a dual-use character. Dual use means on the one hand serving industry usefully, biotechnology or consumer's markets and on the other hand a dangerous weapon in the hands of malicious parties.

It is clear that in view of the many variables in terms of perpetrators, the broad range of weapons, the "weapons" and the usable targets, the number of scenarios imaginable is almost infinite. Furthermore, especially if the threat level is high (for example if there were attacks earlier), a fake attack or an assumed attack (hoax) without physical impact can still have an effect, such as evacuation, panic and also even higher order effects. It also applies that for instance environmental incidents could closely resemble chemical attacks of some kind and vice versa, a form of "crosstalk". In other words, the CBRN terrorism domain is partly related to the safety, health and environment domain.

3.1 Non-State Actors

The chemical threat space has historically shown itself in terms of state-on-state conflicts, however, there is a growing concern about non-state actors using these kinds of weapons. There are indications that various non-state groups considered to carry out attacks using chemical weapons and in fact some of them executed attacks. In recent decades, there are only a handful of prominent cases [10], including the 1995 sarin attack on the Tokyo subway in which a dozen people died and hundreds were injured when the deadly gas was released in five different trains during morning rush hour [11]. In Syria claims have been made regarding rebels using chemicals, including sarin, as well as the Syrian government [12]. Also Da'esh, variously known as IS or 'Islamic State', has been accused of using chemical weapons on several occasions, including chlorine and mustard gas against Kurdish fighters on a number of occasions [13, 14]. Regardless of these clear examples it is still argued that the perception of the probability of chemical attacks by non-state actors may be different from the reality as a result of sensationalism or fear.

Similar to chemical weapons attacks, chemical accidents may also cause enormous destruction and create thousands of victims. Attacks on chemical facilities which ultimately lead to the release of toxic materials do fall within the scope of this chapter. In July 2015, two explosions at a petrochemical plant took place that were believed to be a malicious act due to the distance between the two tanks [15]. "Such attacks are not only within the realm of possibility but are also within the reach of most non-state groups" [16].

In Fig. 1 the several steps of the development, use and consequences of a chemical weapon are shown in chronological order: motivation and intent, development of chemical weapon capabilities, and consequences and effects.

Regarding ideology and motivation there is a paucity of studies that analyse why groups would want to acquire and use CBRN weapons. In general, due to the lack of (statistical) studies, empirical analysis of CBRN attacks is virtually impossible and it is difficult to comprehend the potential extent of attacks by non-state groups using CBRN weapons [17]. Also, the possibility of "lone wolves" conducting chemical attacks cannot be ruled out. For instance, in the biothreat area there are strong suspicions that the anthrax letters case of 2001 in the USA was due to the actions of a single person with access to a US biological defence laboratory [18]. The fact that several sophisticated steps need to be taken before chemical weapons can be developed, makes it unlikely that this could be accomplished by one single person. Rather,

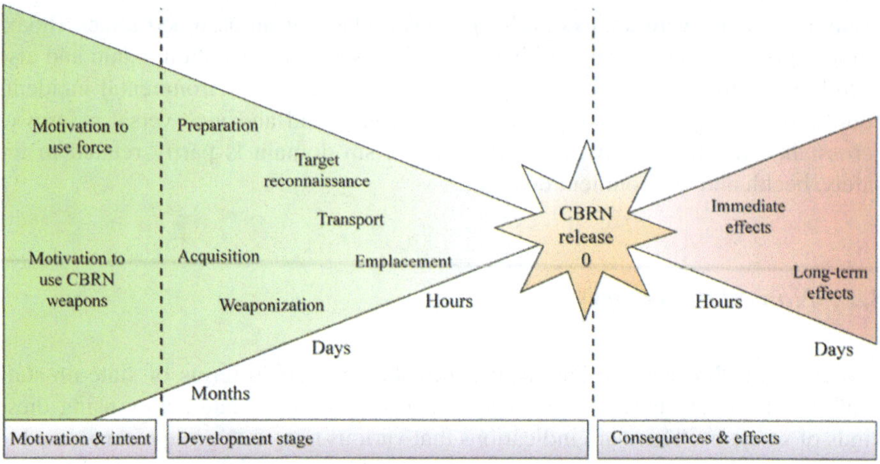

Fig. 1 Overview of the various "building blocks" of the threat: from motivation and intent through actual completion and triggering of a chemical weapon to the impact (*Credit* M.S. Nieuwenhuizen)

one requires a network approach; individuals with dedicated tasks, leaders, financers, suppliers, transporters, bomb builders, those who plant the weapons, triggermen and those who exploit the attack. Research has shown that networks are used not only to recruit, train and prepare for an attack, but also to compensate for elements that a single person may lack, such as resources, elite access or ideological support [19].

Ideology plays a decisive role in a group's objectives and modus operandi. The use of chemical weapons may or may not fit a non-state actor's ideological agenda or encompass overall and specific operational objectives. Based in part on previous chemical incidents, several ideological groups have been identified as groups that may use violence, including possibly chemical weapons.

The following types of groups are distinguished:

(a) ***Nationalist, separatist or irredentist groups***: These are organizations whose purpose is focused on achieving some types of political objective. The prospect of chemical weapons use in areas where the group's supporters are not present is conceivable, but nationalist/separatist groups will generally avoid these weapons because of concerns about alienating their constituencies.
(b) ***Radical religious groups***: These groups are comprised of religious extremists who embrace political objectives and/or forcibly insert religion into the political sphere. Constraints against using chemical weapons as a means to achieve their goals may be overcome by radical interpretation of their respective religious laws.
(c) ***Apocalyptic millenarian groups or "new religious movements"***: These groups show certain similarities with radical religious fundamentalist groups. They aspire to bring about Armageddon, or world destruction. Leaders of such groups may reason that the use of chemical weapons can trigger the apocalypse, thus, in their view, cleansing the world of evil. For example, the Aum Shinrikyo cult

had conducted chemical weapon attacks, the 1995 sarin attack on the Tokyo subway system.

(d) ***Single-issue groups***: They tend to focus on very specific or relatively narrowly defined causes of various sorts, e.g. anti-abortion groups, eco-groups and animal rights activists. It is unlikely that these groups will use chemical weapons to achieve their goals, as indiscriminate weapons may target too wide a spectrum of victims outside their targets, which may adversely affect public acceptance of their agendas. An odd sub-category is "green anarchists," who apply similar perspectives as social anarchists but also critique the way humans interact with the non-human world (animals, nature, etc.). This group shows strong similarities to apocalyptic groups. The possibility of these single-issue actors such as these using chemical weapons thus cannot be ruled out.

(e) ***Right-wing groups***: These groups seek to restore national greatness, suppress "dissident" opponents, expel or subordinate ethnic and cultural minorities or overthrow the existing democratic and "plutocratic" capitalist systems to establish a revolutionary "new order." Individuals in this category are a significant threat for low level chemical attacks, but probably do not represent a threat for mass-casualty chemical attacks due to resource limitations.

3.2 Non-State Actors: Discussion

The fear factor may convince non-state actors to consider using chemical weapons. They may have similar fears about possible contamination caused by chemical agents that beset the general populace themselves. This may deter non-state actors from pursuing use of chemical weapons. They are aware that, during the development and weaponization phases, the handling of toxic and super toxic materials poses significant risks.

Also the fear of severe retaliation may deter a group using CW. However, this restraint does not apply to transnational groups that are spread all over the globe and do not depend for their survival on their continued presence in, or control over, specific territories. If a non-state group believes that it can conceal its involvement in a chemical attack, it may not fear retaliation. Non-state actors may also consider the negative impact on their support systems of using chemical weapons for example their funding sources being affected. Also, the type of action carried out may go against their constituents (support base) moral beliefs. The use of toxic materials, if it has the potential to inflict mass casualties, likely falls into this category for most non-state groups and their supporters. This type of action may be difficult to justify and therefore, rather than impress or inspire, may harm their cause in terms of financial support and recruitment. Ultimately, this could lead to a group's demise. Only the most fanatical or desperate extremists will fail to recognize the broader negative impact that their violent actions are likely to have on their cause [20]. Observations in the rise of IS in the West, suggests that these broader negative impacts were in fact sought after and used to raise money in the early days of that movement, from otherwise blameless individuals in Arab countries, playing on decades of grievances and referencing the colonial era feelings in national inadequacy and anger from those times. As for the topic of suicide actions, the question arises as to why hardly any suicide chemical attacks have taken place, although the chlorine bombings in Iraq from 2006 to 2007 may belong to this category [21].

3.2.1 Probability of Chemical Attacks Perpetrated by Non-State Actors

The probability of a threat being carried out is often determined by the motivation of the perpetrator and their capabilities and resources. The motives behind why a group would choose to develop and use a particular chemical agent not only depends on the ideology, objectives and characteristics of the group, but also on the context. With respect to capabilities significantly fewer hurdles are foreseen for non-state actors to chemical weapons acquisition as a result of increasing availability of knowledge, techniques and dual use materials [22].

To carry out chemical attacks, a non-state actor needs to possess or acquire certain capabilities. Many activities need to be conducted, supported by personnel and resources. One needs to acquire the knowledge regarding materials and how to handle them, financial means to obtain these, the support of suppliers, and the knowledge and capabilities to manufacture the agents and the weapons. The required specific expertise and skills are not likely to be found in one single person. Small networks of persons are likely to be established, which will likely include several individuals with dedicated tasks. To assemble such a network of skilled individuals, recruitment activities must be performed using social media platforms and targeting groups who may be supportive underpinned. Non-state actors derive income from a variety of sources, often combining both lawful and unlawful funding to underpin their activities. One can observe this in the conflict areas in and around Iraq and Syria.

Due to the complexity of producing chemical weapons and the fact that preparing, transporting and employing these weapons involves safety risks, non-state actors may want to install safety precautions such as physical means of preventing chemical materials from being released unwillingly or of protecting themselves against exposure to such materials. Such measures may attract attention in public locations and therefore covert operations may be the result [23].

3.2.2 Production Processes

Production of agents is a first major step leading up to a chemical attack. The lack of governmental control may allow for ideal locations for this. Aum Shinrikyo, for example, managed to operate a state-of-the-art, secret laboratory at the base of Mount Fuji [24], and Al Qaeda reportedly maintained chemical weapons related facilities in Afghanistan [25]. Allegedly, Da'esh has also set up a special branch to develop chemicals using scientists from Iraq and Syria [26]. Extremely covert settings are not always a prerequisite for developing a chemical weapon. In 2005 [27] a British court convicted an Algerian for planning to use poisons including ricin and cyanide in his flat in north London. In June 2018 [28], German security services uncovered an alleged terrorist plot involving ricin by a Tunisian extremist in Cologne inspired by Da'esh. On a larger scale, unscrupulous suppliers have consistently flouted international restrictions on controlled dual use items [29]. It is extremely difficult to shut down such supply lines as export controls based on national or international non-proliferation can be undermined. For example, a Dutch entrepreneur supplied large amounts of raw materials that can be used as precursors for chemical weapons. A Dutch court ruled that he was complicit in a violation of humanitarian law by supplying substantial amounts of raw materials for mustard gas, and he was sentenced to seventeen years imprisonment [30]. In 2013 [31] it was confirmed that European companies had exported a number of different dual use chemicals to Syria, again demonstrating that there are inherent difficulties in controlling the international transfer of chemicals.

3.2.3 Means of Delivery

These are understood as the equipment to transfer chemical agents to the target zones or to release the agents to the vectors (air, water, foodstuff, animals or plants). In general, means of delivery, with varying technical sophistication and effectiveness, can be mechanical or pneumatic devices to disperse the agent, or explosive dissemination, military devices to carry the agent to the target zone (rockets, missiles), or "ordinary" transport vehicles (aircraft, crafts, trucks, cars, trains, parcels and letters). The most important factor in the effectiveness of chemical weapons is the efficiency of dissemination. In principal chemical agents can be dispersed as a gas (vapour), liquid, aerosol (mist) or as solid aerosol (smoke).

Obtaining suitable delivery systems can be a problem, although covert attacks can be carried out from fixed locations in urban areas. Suitable dual use delivery systems are readily available, e.g., using crop duster aircraft and simple spray generators that can be readily adopted for a variety of agents. The lethality of any given chemical weapon would also increase strikingly if it were to be released in a closed environment like an office building complex or in a high density urban environment.

With respect to the efficiency of attacks it should be noted that not only the number of casualties is a key parameter, causing public fear and panic, and disturbance of the society may be the main goal of attackers. In such cases, multiple simultaneous small attacks, which can be carried out by use of very simple means of delivery, could be more efficient than one big strike.

Many chemical agents need to be weaponized prior to delivery. After production of a suitable quantity of agent (matching the desired target) it must be stored appropriately to keep in stable form, avoiding evaporation or degradation. Thickeners or carrier agents can be added to liquid compounds to enhance the agents' viscosity and persistence on surfaces or dispersion properties, respectively. The chemical agent must then be inserted into appropriate munitions or dissemination devices to be used on a target [32].

Spray delivery is one of the most effective methods for dispersing chemical agents, but industrial accidents, explosions, and spills can also release them into the atmosphere. The accident at the Union Carbide's pesticide plant in Bhopal in 1984, which killed 8000 people with at least 150,000 injured, killed far more people than the sarin attack of Aum Shinrikyo in the Tokyo subway system.

For application as vapour or aerosol sophisticated systems with sprayers mounted on aeroplanes can be used. But terrorists could also employ chemical weapons agents in such simple means like commonly available aerosol generators (glass bottles from deodorant sprays), commercial garden sprayers or fire extinguishers.

In a military context munitions include aerial bombs, mines, submunitions, projectiles, warheads, artillery shells and rockets. Some nerve gases are delivered by binary weapons, where two less toxic substances are enclosed in a delivery vehicle e.g., an artillery shell. The shock of firing the munition ruptures the seal separating the two substances allowing them to mix and form the lethal agent such as sarin.

Dispersion of chemical agents by explosives is less effective because many chemicals are flammable and partially ignite. Moreover complex chemical molecules are sensitive to high temperatures and can deteriorate if exposure is too lengthy.

Other factors that affect the successful delivery of chemical agents include:

- **Environmental conditions in the target area**—higher air and ground temperatures cause faster evaporation
- **High humidity**—can cause larger droplet sizes reducing the effectiveness but increasing the persistence
- **Precipitation**—heavy rain dilutes and disperses chemicals; snow can enhance the persistence
- **Wind speed**—wind can disperse vapours, aerosols, and liquids, thus enlarging the affected target zone and diluting the concentration in air
- **Building construction**—buildings can absorb/adsorb agents, but also offer protection
- **Nature of terrain**—woodlands and hills create greater turbulence in air masses and can cause enhanced deposition of contaminated aerosol particles.

For a terrorist attack in closed buildings tanker trucks could be used to deliver the chemicals needed. Some lethal gases or chemicals rise (lighter than air), but all current chemical weapons are actually vapours that are heavier than air. This fact can be exploited by terrorist attackers in buildings with large open spaces or forced air systems, or by using elevator shafts and other vertical corridors.

3.2.4 Executing the Attack

Execution of the attack is the final step. Creating and employing systems capable of inflicting massive physical impact is quite challenging. The scope of the effect of a chemical weapon is largely determined by the nature of the threat agent as well as the weapon system used, in particular the efficiency of its delivery. Chemical weapons can dissipate before harming many people if meteorological conditions are unsuitable for dispersal [33]. Also the purity of the chemical agent has a large impact on the ultimate effects of an attack. In the sarin attack on the Tokyo subway system, for example, the full potential for catastrophic damage was not achieved as the sarin that was used was not pure and furthermore the dispersal technique was not well developed·. To date, methods used by non-state actors to disperse chemical agents have been primitive and inefficient [34]. What seems likely in the future, although still not a simple endeavour, is the creation of a "dirty bomb," combining components that are accessible and not necessarily expensive with an improvised explosive device. Once a target has been identified and meets the actors aims then a more detailed plan would be put in place and rehearsed. Finally, a triggering/detonation mechanism, set off either by a triggerman on the spot or via a remote-control system, would be necessary to ensure release of the materials. The non-state actor may lack control over the effects of chemical agents after the release, particularly if the actor has developed a primitive device in which low grade toxic materials are used or higher-grade materials are improperly disseminated.

4 State Actors

There is a long history of state chemical warfare dating back to the Chinese use [35] of smokes laden with arsenic, about 1000 years BCE and in the Hindu epics such as the Mahabharata (or Fifth Veda, dating back to 400 BCE) in which chemicals were used to poison water supplies (the concept of '*just war*' forbade their use on arrows [36]). It is certain that chemicals would have been used before these dates, though the notion of 'state' becomes increasingly nuanced as we go back in time.

Indeed, while chemical warfare agents were made illegal after their sustained use [5] in World War I there have been a succession of agreements and treaties which prohibit their use.

For example, in 1675, the Strasbourg Agreement between France and the Holy Roman Empire sought to prohibit the use of poisoned bullets. The Hague Convention of 1899 in article IV 2, forbade the use of projectiles whose sole object was to carry asphyxiating gases. The Geneva Protocol [37] was signed in 1925 and prohibited the use of poisonous gases.

The use of CWAs by state actors continued before and during World War II. Within 3 years of the signing of the Geneva Protocol, the Italians used mustard gas extensively in Libya and later in Abyssinia (now Ethiopia). However, it is important to note the use of Zyklon B, hydrogen cyanide, tabun and carbon monoxide by Nazi Germany in killing 6 million civilians, by far the largest use of CWAs in history. The Nazis also used CWAs in Crimea [38] and probably elsewhere.

In 1943, the US ship SS John Harvey was sunk off Bari, by German bombers with mustard gas on board [39]. The end of World War II did not stop widespread and again illegal use: for example it is claimed [40] that the Rhodesians killed many, probably thousands, of people with organophosphates in the 1970s. The Iraqis used a range of CWAs against the Iranians and others from 1983 onwards and in the Gulf War [41], killing far more than 100,000 people.

State-sponsored use of CWAs is not always admitted, but it is also arguably used on a denied basis to make a political point. It has been suggested that Alexei Navalny, a Russian/Ukrainian politician, was poisoned, according to German Authorities, by Novichok nerve agent, a cholinesterase inhibitor.

What is perhaps the most intriguing part of this incident is the choice of Novichok as the weapon, a chemical intimately associated with Russia and with a history of use against many targets. Its choice seems designed to send a state-sponsored message with arguably global ramifications [42]. It is instructive to revisit the Skripal incident, which took place in 2018 in Salisbury (United Kingdom). Former Russian military intelligence (GRU) officer and Sergei Skripal [43] and his daughter were poisoned with Novichok. They were fortunate to have survived, as did an attending UK Police Officer, who was also contaminated. However, one unconnected person died a few months later after finding a perfume bottle that had been discarded but presumably had been used in the original attack. The two Russians implicated in the poisoning of Sergei Skripal arguably left a 'bread crumb' trail for the purpose of sending a message [44].

5 Impact of Chemical Warfare Attacks

Although in many cases it is extremely difficult, or not possible at all, to conduct a sound quantitative assessment to characterize the possible consequences of a chemical attack, attempts have been made to rationalize the impact of the attacks in terms of risks and vulnerability [45]. A framework of definitions is an element of that rationalization approach:

(a) **Risk** of chemical attacks to society is a combination of vulnerability and the likelihood that such an attack would occur, i.e. the likelihood that a terrorist would use a certain quantity of some type of chemical weapons for carrying out his intentions.
(b) **Vulnerability** of society to chemical terrorism is the damage (in terms of physical and psychological health, societal functioning and the economic and political viability of a country) incurred in a chemical attack (including hoaxes).
(c) **Health** includes the physical and psychological health of the population, crops and livestock.
(d) **Societal functions** are those services to a population that need to be delivered on a regular basis in order to ensure both physical survival of a population and the proper functioning of society. Societal functions are produced and delivered by people, procedures, critical infrastructure, fleets and mobile equipment of various sorts, raw and finished materials. Critical infrastructure is the name for the assortment of sites, buildings, immobile equipment and connections between these that are used for providing societal functions.

(e) **Economic viability** of a country is the unhindered macro-economic development of a national economy encased in the stock market, investment climate, and labour market.

(f) **Political stability** (viability) of a country is the resilience of a country to political shock. Thus, the risk of chemical terrorism is a subset of the overall risk to society. This overall risk also includes the openness of a country to the influx of international terrorists, the presence of home-grown terrorist groups, the likeliness that a person becomes involved in a group intent on terrorism, receives his training, and develops the definite intention of carrying out a terrorist attack.

(g) **Countermeasures** are aimed to reduce the impact of a chemical attack would be a mix of actions deliberately implemented by government to prevent terrorists acts from taking place and/or to prevent a terrorist act from causing damage, including technical means such as directed medical countermeasures and other means such as planning, awareness raising and exercising.

(h) **Efficacy** of a mix of countermeasures against chemical terrorism is the effectiveness of this mix in limiting the likelihood that a chemical attack would occur and, would one occur nevertheless, of limiting the damages of a complete set of plausible terrorist attacks to health, societal functions and the economic and political viability of a country.

(i) **Efficiency** of a mix of countermeasures against chemical terrorism is the comparative cost (in any cost accounting terms, including human health, societal function, political stability and monetary) of this mix gauged against a standardized set of plausible terrorist attacks and a standardized efficacy of a mix of countermeasures.

5.1 Tiers of Impact

It is on the above definitions that the following model of society and the impact of chemical terrorism are based. As to effects/impacts of a chemical terrorist attack, the following three tiers are defined (and shown in Fig. 2).

- Tier 1: Impacts on health and attendant first responders' actions constitute the first-tier effects.
- Tier 2: Impact on societal functions constitute the second-tier effects.
- Tier 3: Political and economic impacts.

It should be noted that in these tiers no temporal ingredient is present. Usual one could expect that the order of duration would be tier 1 < tier 2 < tier 3, but in specific cases a tier 1 effect might well exist for a longer period than a tier 3 effect.

Second tier effects are largely caused by first tier effects, and third-tier effects are caused by second and first tier effects. The root cause of a third-tier effect must be imported from a damaged societal function in the second tier of the model, and connected to this, from a first-tier health effect. First tier effects (health) are the result of the contamination actions carried out in the '0-th tier effects circle', that is,

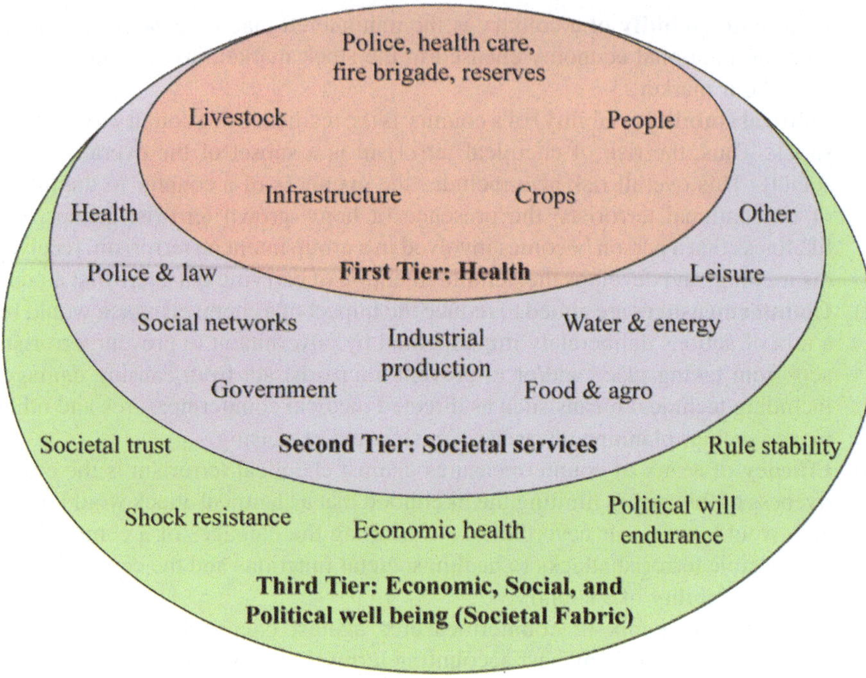

Fig. 2 Model of impact areas in Society (*Credit* M.S. Nieuwenhuizen)

the terrorist action. The vehicle here is the chemical contaminant. The spreading of first to second tier effects requires vehicles like quarantine policy measures, news of a chemical attack on the media, people fleeing the scene of an attack. A third-tier effect may be conveyed from a second-tier effect through economic repercussions of breaches in societal functions, causing investors to lose faith with the political acumen of a government, causing stock markets to fail etc. The complex interlinkages that make up modern society however also provide feedback loops and 'fast forward' loops (i.e., a direct political effect after the assassination of a single person) between the tiers of the model.

Only in the totality of first, second and third tier effects can the final outcome of a terrorist attack be known. Likewise, only in the working of the mix of countermeasures against all these effects will its efficacy be shown. The terrorist impact damage mechanism thus described is shown in Fig. 3.

Modern society is a complex, intertwined system, in which a small disturbance in one area may have a disproportional effect on the system as a whole. In fact, the system character of modern society may imply that certain types of attack could cause the system itself to lose stability, i.e., the effect of a large-scale chemical attack might grow exponentially out of control because contaminated people move around looking for medical aid thus affecting ever more people so that the health functions

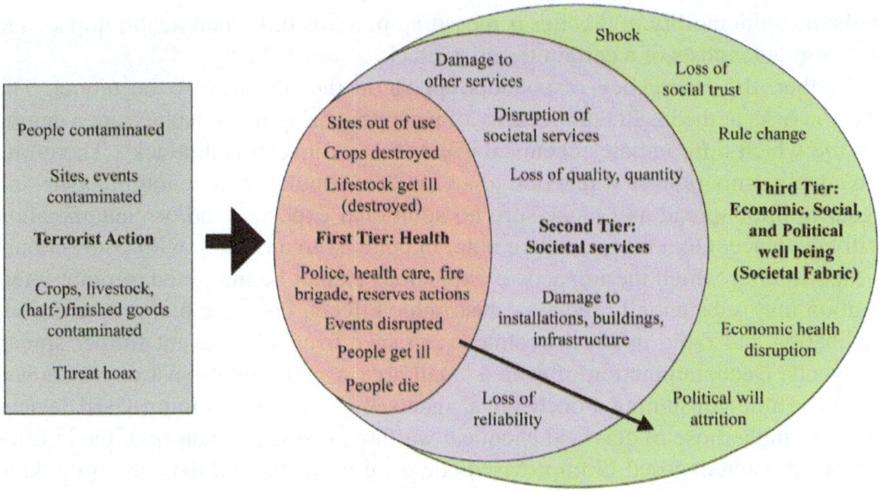

Fig. 3 Schematic view of Terrorist impact mechanisms (*Credit* M.S. Nieuwenhuizen)

find themselves unable to cope and so that an ever-increasing number of societal functions are disrupted.

Buildings and/or infrastructure may be damaged, which constitute additional first tier impacts of an economic nature. Where the costs associated with these damages would be higher, vulnerability would be greater. A well-protected population with ready access to decontamination and inoculation would be less vulnerable than one without. There will be costs to be met for stopping the terrorist event, stopping its direct impact (mitigation), delivering medical aid to victims, policing the site of an event, putting out fires, etc. These costs are to be included into the damages that a society incurs as a result of a terrorist attack, even though high costs here resulting in a lower loss of life may be greatly preferred to the other way around. It thus presents us with an effectiveness/efficiency question. Vulnerability in this tier is measured in terms of human health impact and first responder costs of a certain terrorist attack.

5.1.1 The First Tier of Attack

At the **first tier** those elements of society are found that are directly involved in the health effects of a terrorist attack. During a chemical attack, people and the environment are exposed to and may be contaminated by toxic materials. After such attacks, the primary immediate need is to save lives and treat the injured. In the model the preferred vehicles of a chemical attack are taken into consideration, i.e., food and sites/events. Buildings and/or infrastructure may be damaged, which constitute additional first tier impacts of an economic nature. A well protected population with ready access to decontamination and inoculation would be less vulnerable than one without. Vulnerability in this tier is measured in terms of human health impact and first responder costs of a certain terrorist attack.

Further, the emergency responders appear in the first tier of the model. The specific risks to the health and security of personnel bringing assistance are a unique feature which differentiates chemical events from conventional attacks. Therefore, the equally important first reaction to an incident should be to contain the site and agent to limit spread and make sure no additional exposure and/or contamination will take place. On a more positive note, in contrast to an attack with conventional weapons, after which the impacts of explosions cannot be mitigated, prompt intervention may substantially reduce catastrophic effects. For example, proper decontamination and rapid medical treatment can save lives and prevent further spread of agents. Decontamination of even a small area would require vast resources and would be time-consuming. For chemical agents, the symptoms of improvised devices may resemble those of classical chemical warfare agents. The nature of the injuries sustained when exposed to improvised chemical weapons will depend on the kind of agent; for example, the agent may attack the body's central nervous system, like nerve agents such as sarin or VX. Health facilities may be paralyzed by the arrival of people affected by chemical agents. Hospital staff may fear secondary exposure (even though the dangers of secondary exposure are limited, especially in the case of non-persistent agents) and may thus be reluctant to treat patients without wearing protective equipment. Attacks with improvised weapons are less likely to cause many deaths, but a society can nevertheless be gripped by panic once it is known that a chemical weapon has been used. This panic can also be a significant burden on the healthcare system if many of the "worried well" flood hospitals. They may deprive actual victims of getting the medical attention they need. The time from exposure to onset of symptoms is generally relatively short (minutes to hours) for acute doses of chemical agents or toxins. Nevertheless, psychological effects, such as fear of not knowing whether one is contaminated, may largely dominate the immediate effects of a chemical attack.

5.1.2 The Second Tier of Attack

The **second tier** consists of a number of societal functions. Those functions are included that were deemed to be most vital to the survival of its citizens and the proper functioning of society. Second-tier effects are the immediate societal function effects of the sudden emergence of a large group of poisoned persons, foodstuffs, events, buildings, and the emergency responders' activities. They are a consequence of first-tier effects spreading to sectors other than those immediately affected. Second tier effects may disrupt the normal functioning of parts of society. For instance, personnel taking care of vital societal functions may be taken ill and cannot be replaced at short notice. The attention that 'emergency responders' would be required to deal with the aftermath of a chemical attack could make them incapable of taking care of their normal functions. Societal functions like the supply of water, food, electricity and gas, sewage and rubbish collection functions, health and security provision, communications or the flow of agricultural products into the food chain can be substantially affected. Similarly, key installations and buildings (for instance, hospitals) may be contaminated and therefore taken out of action.

The health and first response impacts taken together might hinder the smooth functioning of parts of society. Some vital functions might be interrupted such as the sufficient delivery of water, food, shelter, electricity, gas, etc. These constitute the second-tier effects of terrorism. The costs of such a serious breach of a societal function might be much higher than the direct health costs of a terrorist attack.

It is even possible that a terrorist attack would be directly aimed at these functions. The present worldwide attention lavished on critical infrastructures is predicated on this point. It bears pointing out, however, that evidence thus far suggests that terrorists' actions are mainly aimed at hurting people or attacking sites or buildings that contain some powerful symbolic value (for instance, the military power of the state, or religious buildings). They have, thus far, shown a remarkable reluctance to comply with the target categories suggested by 'protection of vital infrastructures' advocates. In fact, the theme of 'vital infrastructures' seems to be connected to the American 'brittle infrastructures' debate of the 80's rather than to any actual terrorist target or goal. Nevertheless, it may be a good idea to have a close look at the interconnections between infrastructures, if only to reduce vulnerabilities to natural disasters.

Vulnerability on these aspects (that is, to societal impacts) is measured in terms of the costs associated with breaches in the provision of societal functions for a certain terrorist attack.

5.1.3 The Third Tier of Attack

The **third tier** are overall indirect effects in all sectors of society and consist of the psychosocial, economic and political viability of society. The economic [46, 47] viability is a singular item (economic health). Infrastructures or facilities may be persistently affected, social norms and rules may change. Trust in the government and the state might diminish over the long term, economic performance might deteriorate and stabilise at a lower level than before, some geographical areas might become unable to sustain the workings of a modern society. These effects would be more difficult to address than first and second tier effects. As serious as first level effects may be, it is at the second and third levels that society as a system is affected in several ways:

(a) **Shock**: Terrorists will generally aim at people since it is usually their intention to spread fright. As a result of first and second tier impacts a sudden and drastic reversal of policy may be decided upon, or the body politic of a country may find itself paralyzed. Any reaction that would normally be unnatural could be described under the term 'shock.'
(b) **Political impact**: Terrorist attacks will generally have some kind of long-term political influence. This is an important one, not least because for the terrorists

this is the most important effects level. Guarding society against this unwholesome influence is possibly the ultimate aim of anti-terrorism measures. In determining the political influence of a terrorist attack, much depends on political and societal circumstances. There is no a priori recipe that formulates the proper way to counter the political influence of a terrorist attack, also because influence is a normal and constitutive part of the political game. In the end the "receiving" political system will have to decide on the appropriate answer to terrorism and thus, on the political influence of it. Where a population feels that no vitally important issues are at stake, or where there is a latent sympathy with known terrorist demands, an attack may well be quite successful.

(c) **Rule change**: The retention of a truly democratic political system along with its attendant societal order and citizen and human rights must be the ultimate goal of any democratically elected government. In a slow and hardly perceptible way, the incidence of terrorist attacks may induce a change in the rules of the political 'game.' Think here of the half-use, half endorsement and not quite condemnation of terrorist threats, attacks and fear for the benefit of an ostentatiously 'legitimate' political party whose goals coincide with the demands of the terrorists. Think also of the political pressure that a diaspora population section or its (self-appointed) representatives may exert on its guest home government. This use of terrorism is in practice much more common in European nations than the spectre usually summoned by left-wing political parties of a 'police state.' It undoubtedly indicates a nefarious change of political rules that should be resisted and rolled back.

(d) **Indirect economic influence**: After the initial shock of an attack, indirect effects apply. The nature of the indirect influence varies with the intensity of the attack, with its kind, and with the response that has been performed. Economic losses stem from the attack itself (people killed or injured also represent economic opportunity losses), from the rendering unserviceable of areas, equipment, buildings or/and infrastructures, and from the loss of economic trust. It is this last type of loss that is most difficult to counter. Infrastructures may be restored, areas cleansed, but trust is extremely hard to rebuild. Lost trust shows in smaller investment, the relocation of business, long-lasting changes in markets, lost share value, and increased economic transaction costs (rents, costs for certain services).

(e) **Psychosocial impact**: Terrorist attacks are often primarily a form of psychological warfare in which a localized incident is intended to spread fear and anxiety among a wider audience or specific society [48]. This psychological aspect is vital to the success of a group. The question is why one would seek to conduct chemical attacks, which are much more complicated. The threat alone of using chemical weapons gives a non-state group many advantages, because the thought of being the victim of a chemical attack spreads greater and even disproportionate fear among potential target groups. The fact that psychological symptoms are more likely to occur when facing chemical agents than when facing conventional weapons is possibly the most important motivational incentive that might make the acquisition and/or use of chemical weapons attractive to non-state actors. Similar to the psychological impact of chemical warfare in World War II, the concept of a chemical weapons' attack today is anxiety provoking not only because of the intangible nature of most of the harmful agents being used, but also because of the doubt as to whether or not one has been exposed. Increased public fear can be anticipated from the possibility of suffering slowly and for a long period or the fact that the effects of chemical weapons may be delayed.

6 Impact Reduction

To deal with the possible impacts of chemical terrorism require a complicated, and preferably coherent, set of countermeasures prior, during and after the attack. This could be defined as a system, often also called security cycle. Several system descriptions have been postulated. For instance in 2009 ESRIF, the European Security Research and Innovation Forum, described [49] a system as depicted in Fig. 4. Later on the EU DECOTESSC1 project [50] depicted the security cycle in a different way, highlighting the 5 elements: threat assessment, prevention, preparation, response, recovery as well as the systems-of-systems concept (Fig. 5).

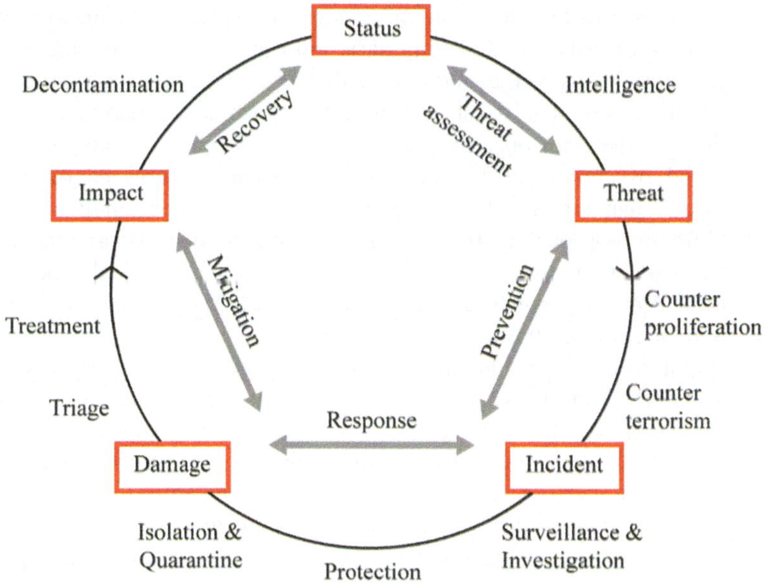

Fig. 4 Security cycle as depicted by ESRIF (Courtesy of ESRIF)

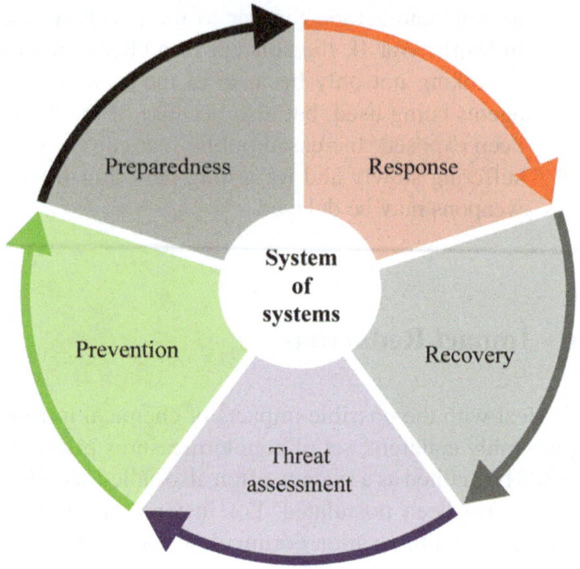

Fig. 5 Security cycle as depicted by the DECOTESSC1 project. (Courtesy of DECOTESSC1)

6.1 The Threat

In an ideal situation the research and assessment activities on chemical risks and threats are comprehensive, multidisciplinary, innovative, up to date and perfectly coordinated—national as well as international. The object of such activities are the organization structures of terrorist networks, their technical capabilities and their methods. It comprises not only the mere feasibility of chemical attacks but also political, social, psychological, economical intentions and coherences; furthermore, the impact on human health and environment. Therefore, experts on different levels and with different approaches, be it theoretical or experimental orientated, work together: scientists with more general models and concepts, technical experts, intelligence, political analysts, police experts and many more. At the same time the needed nondisclosure and secrecy has to be guaranteed.

A key point would be a so-called "shared awareness" of the threat situation, meaning a shared picture which would involve the general public, trade actors and industry, among others. This would ensure a broad and integral approach to potential threats. Rising awareness among the population concerning prevention is regarded as especially important, starting with the first signals when a potential terrorist purchases weaponry or agents. This could be noticeable for the industry or other civilian entities. However, there is disagreement about the sense of raising awareness of the population prior to an incident. The public may be more alert, but this would also go along with increased states of fear. In an ideal situation, the public would be informed about an incident and could look for help themselves. Their self-reliance may even be increased by additional information gained by mass and social media. This view is regarded as too optimistic by some experts who expect that in the case of an incident people would tend to proceed to a place of danger and would be difficult to direct.

The authorities responsible for threat assessment have to share their acquired results about the threat situation with the authorities and companies responsible for prevention, preparedness, response and recovery measures. A strong working relationship between intelligence, law enforcement, public health and other competent authorities is therefore essential to responding effectively to a terrorist incident or deliberate malevolent act. Only if they all have the access to the results of threat assessment can they prepare themselves adequately or perform their work. For achieving this it is essential to establish communication mechanisms between all responsible authorities. These mechanisms and the criteria used to prompt information exchange should be developed with consideration for example of pertinent laws and regulations protecting both the sensitive data of law enforcement and confidential medical information. Furthermore, international cooperation is essential for threat assessment authorities. Instead of 'need to know,' international cooperation should be based on 'need to share' to come to an optimal situation in which all necessary information is shared for a clear understanding of threats.

The threat assessment process needs to continuously evaluate the additional information derived from public health, law enforcement, competent authorities, and intelligence sources. Management level law enforcement, competent authorities and public health officials should begin coordinating as soon as possible. Such coordination enables the implementation of appropriate measures to protect and treat law enforcement personnel who are exposed to suspect material at the scene and elsewhere, as well as to protect and treat the public. If circumstances warrant suspicion that the incident is intentional, law enforcement services will focus their efforts and resources on conducting the criminal investigation as the lead agency.

6.2 Prevention

Prevention measures have mostly a regulatory aspect for what concerns the reduction of the likelihood that a possible terrorist gets know how, material, components, delivery means and everything else needed to attempt and to successfully achieve a chemical attack. Political measures can also be considered prevention measures as far as they aim to solve or attenuate the social, ethnical, or religious conflicts that motivate terrorists.

Generally speaking measures for the protection of material are easier to implement and more effective than protection of intangible components like knowledge. Nevertheless protection of material may be unsuccessful in cases where a device can be assembled in a cellar using ordinary ingredients that can be purchased in a supermarket, as may be the case for some types of explosives.

In the current system, on top of the specific measures in place in individual countries, there is a legal international framework that regulates the utilisation and protection of chemical materials. These international instruments have been implemented in the different states on national laws, regulations and decrees. Important international legal instruments are the 1997 Chemical Weapon Convention (CWC) and the 2006 Regulation concerning the registration, evaluation, authorisation and restriction of chemicals (REACH) in the EU.

Terrorist organisations frequently have an international diffusion. This means that often materials, components, people needed for an attack must travel across borders and therefore adequate border controls on people and goods transit must be implemented. Border controls need to be complemented by an efficient regime of export control of sensitive technologies, materials and knowledge.

As stated in the previous section, many of the prevention measures have an international framework, so there is overlapping/sharing of responsibilities between national and international organisations. Generally speaking the national organisations have the duty of implementing the prevention measures, while the international organisations maintain a verification regime through inspections. Most of the international treaties and conventions have an international organisation in charge of the verification regime: for the CWC it is the OPCW. If, notwithstanding the above described prevention measures, the chemicals has been diverted from the legal protected location and become available to potential terrorists through black market or other clandestine networks, then the responsibility to retrieve or intercept it before the attack falls within the responsibilities of the traditional law enforcement bodies: police, border control forces, security organisations, civil or military intelligence, secret services and others. A peculiar aspect of chemical threats is the generally high number of national authorities and organisations involved in the different aspects of prevention, response and crisis management. A clear definition of roles and responsibilities is needed in order to avoid confusion, conflicts or omissions during an accident.

Private industry dealing with threat material has the responsibility to maintain an adequate level of security and the obligation to report immediately any accident, release, theft or loss of control of material. A possible danger can be derived by the fact that an industry subject to an accident involving loss of chemicals could tend to hide the accident in order not to damage its image or incur economic losses; such behaviour will fatally delay the intervention of law enforcement decreasing the capability of the system to react. In addition private industry needs to acquire and develop a major awareness and sensibility towards dual-use technology and equipment.

The role of the population should not be disregarded and can play an important role when properly informed (public awareness). Raising awareness of the population in the context of prevention is important, because the first signals of terrorists buying weaponry or agents are noticeable for the industry or other civilian entities. Nevertheless the point concerning how much the public is to be informed is frequently controversial, in particular pre-information on the risk prior to an incident. While the population might be more alert, there could be the possibility to increase fear and as a result the population can panic when the incident happens. There is a tendency to disagree on the fact that not only too much information but also specific information to create awareness will lead to fear among the population. More or less the same line of reasoning was observed for resilience and self-reliance of the population.

6.3 Preparedness

The ideal measures to handle a chemical threat depend partly on the threat itself but to a large extent the main procedures follow the points below, divided into measures regarding procedures and more technical measures.

Procedural measures involve chemical threat identification, risk assessments, creation of relevant databases, defining action plans, and instructions to both first responder teams and the public. A key part of this is developing a communication plan and the effective use of media along with appropriate training, education and exercising.

Technical measures involve simulation and prediction of the possible incident and the hazard zones, implementation of a real-time warning and/or surveillance, introduction of stand-off detection systems and easy to use point/handheld detection systems.

The responsibility for preparedness varies depending on the type of threat and the type of object which is being in focus. The organisation and responsibility also vary between countries including both civil and military bodies. Typically, the authorities such as public health, food safety, radiation safety authorities, and private actors are responsible for the preparedness of their own areas but have been given that responsibility from higher authority handling civil contingency and rescue. Above them, the national government have the ultimate responsibility.

6.4 Response

Response phase is related to the actions which prevent the effects of an attack from spreading like immediate awareness of nature and severity of the incident, activation of responsive measures, search & rescue, isolation of people, identification systems, diagnoses and training of first responders. Well-developed emergency response plans designed for emergencies involving hazardous materials are the foundation of an efficient response. The objectives of a response plan would include coordinated response, definition of roles and responsibilities, minimizing the impact of incidents or attacks, prevention of loss of valuable classical forensic evidence, performing forensic investigation and support prosecution, as well as reporting and dissemination of information.

In the case of a chemical threat the marching order that describes the situation is: threat and danger estimation, isolation, evacuation, threat and danger re-estimation, and prevention of additional damages.

After this it is time to focus what are the actions to be made in each situation. For chemical incidents sufficient reactions and a fast capability for alerting the organisation is followed by the first responders protecting themselves after the arrival at the location, recognising the situation (fire fighters, police), determining the danger zone (fire fighters, radiation safety authorities, meteorological institutes), its isolation (police) and warning the civilians (fire fighters emergency centre). This is followed by mitigating the impacts of the hazardous substance or its elimination (fire fighters and police), decontamination and first aid of the victims (fire fighters, emergency medical service), the monitoring of the victims´ condition (health care) and predicting the spread of the hazardous substance (fire fighters, meteorological institutes, radiation authorities). After this comes the risk analysis (fire fighters, police, experts) and criminal investigations (police), collecting evidence and their analysis (fire fighters, police, criminal investigation laboratories) and informing people (fire fighters, police) of what has happened. Giving out information to the public is handled by the highest responsible person, or a person appointed by him. Who that is depends on the level of the incident and can vary from a police officer to the prime minister.

The current doctrine in mitigating damages after a large-scale chemical emergency is mostly focusing on cases where the exposure of victims to chemical agents is through inhalation or contact with the agents. The procedures consist of triage, mass decontamination, first aid and medical care, if necessary. Chemical incidents may also require infrastructure and environmental decontamination before returning back to normal use.

Responding to a chemical emergency may require considerable coordination internally within a state as well as externally with neighbours and partners involving a broad range of governmental and nongovernmental stakeholders. The probable incident starts with alerting the police, the fire department and the ambulance services at the location of the emergency with uncertain information of what has happened. The first stage of operations and management are handled with existing resources and personnel. The ideal way would be an information-based start with aimed resources, but this rarely happens in the real world.

When the situation has been assured the possible operatives consists of for example the police (counterterrorist police, criminal police, criminal investigation laboratory), the rescue department (defence forces, technical research centre of defence forces). Other participants may include meteorological institutes, health care districts and transport departments. To warn public, national and local broadcasting companies may be utilised.

While the involvement of military personnel in chemical emergencies is possible in a number of nations, the process is typically not straightforward. In many instances, it may take time to obtain the required political approval, and any military assets employed would at best be assigned to decontamination tasks.

6.5 Recovery

The recovery procedures from a chemical incident depend largely on the type and nature of the disaster: where it happened, what was the target, which agent was used, and whether the attack was covert or overt.

One major long-term impact of a chemical incident is principally the disruption it causes to normal life by denying access to the sites affected by the release. Depending on the target of a chemical attack, decontamination may be needed along the food production chain, water supply systems, buildings and other areas. A well-managed recovery strategy can significantly reduce the overall impact of the incident, especially where the critical infrastructure is affected. After decontamination, the waste has to be neutralised or stored at monitored waste treatment plants or waste storage places.

In general, the healthcare system should be equipped to treat the mass casualties that might arise from a chemical attack. The healthcare system should be able to rapidly enhance the capacity to adequately care for large numbers of patients (surge capacity) and victims with unusual or highly specialized medical needs (surge capability). Especially for infectious patients there has to be a sufficient number of isolation rooms available. In an ideal situation there should also be plans for alternative patient care sites and a possibility to quickly create temporary isolation enclosures for infectious patients.

A CBRN incident is likely to involve a number of national ministries and departments within the nation. The actual configuration of the participating departments depends on the type of attack.

7 System of Systems

Maintaining security against chemical threats presents several challenges for society. Multi agency response to chemical incidents is by their nature dynamic and confusing where accurate information about the event is sparse. It is imperative for incident managers to join their forces early in a possible chemical event to limit the casualties and community disruption. Given the large number of possible targets and ways of causing widespread injury, it is necessary to have a comprehensive, integrated and collaborative approach to counter these threats.

A system of systems security ideally allows the optimal use of available expertise and resources in the most efficient manner so that different actors are linked together in order that they can exchange information to achieve shared awareness and coordinate their actions potentially increasing speed of execution. Such a system might produce significant efficiencies in terms of sharing skills, knowledge and limited resources as well as gaining the synergy of providing a common operating picture and being readily able to share information. Ideally, such a system also includes the public and other actors such as Non-Governmental Organisations (NGO) and private companies.

In an ideal system, information about the incident is disseminated to all the parts at a correct level and without unnecessary time delay. Optimum use is made of different media channels in order to reach as many as possible of the population, and there are plans for different parts of society including the public on how to respond to the chemical threats.

An important issue in such a security system is the efficient flow of information both vertically within each security cycle phase and horizontally between different phases. This may include both organizational and legislative issues like what information and to whom the information will be made available. In turn, to facilitate integration of detectors and other components architecture is needed, which is based on open and widely accepted standards defining for example the hardware, data formats and communications protocol at the interfaces of the security system.

In order to get an organisation to work, the chain of command needs to be clear and well known to all actors. In the ideal situation the authority of, and the knowledge about the coordinating body is familiar to the involved parties, and the leadership is clear in case of incidents demanding cooperation across national borders or incidents activating the national measures. Training, national and joint between member states, which spans over several phases is essential to get the system of systems to work in chemical countermeasures.

A major issue is also the public acceptance of the security measures. Although the new technologies like closed circuit television with face recognition and suspicious behaviour detection, passenger screening, data retention, eavesdropping etc. are expected to enhance public security, they are subjecting ordinary citizens to an increasing amount of permanent surveillance, potentially causing public fears about infringements of privacy and civil liberties.

References

1. Paracelsus (1538) Dritte defensio
2. Romano JA, Lukey BJ, Salem H (eds) (2008) Chemical warfare agents. In: Chemistry, pharmacology, toxicology, and therapeutics. CRC Press, Boca Raton, pp 1–20
3. Smart JK (2004) History notes: chemical and biological warfare research and development during the Civil War. CBIAC Newsletter 5(2):11–15
4. Owen W (1917) Poem entitled "Dulce et decorum est"
5. Harris R, Paxman J (1982) A higher form of killing: the secret story of chemical and biological warfare. Hill and Wang, New York
6. Sartori M (1939) The war gases: chemistry and analysis. Van Nostrand Inc., New York, pp 3, 33, 59, 165, 181, 188, 217
7. Formally: Agreement on destruction and non-production of chemical weapons and on measures to facilitate the multilateral convention on banning chemical weapons
8. Formally: Convention on the prohibition of the development, production, stockpiling and use of chemical weapons and on their destruction
9. See for example Quran verse 2.190
10. Monterey WMD Terrorism Database provides an overview of worldwide incidents involving the acquisition, possession, threat and use of weapons of mass destruction by sub-state actors, based on open sources. wmddb.miis.edu/. Accessed Sep 2019
11. Pangi R (2002) Consequence management in the 1995 sarin attacks on the Japanese subway system. belfercenter.ksg.harvard.edu/files/consequence_management_in_the_1995_sarin_attacks_on_the_japanese_subway_system.pdf. Accessed Aug 2020
12. McElroy D (2013) UN accuses Syrian rebels of chemical weapons use, The Telegraph. www.telegraph.co.uk/news/worldnews/middleeast/syria/10039672/UN-accuses-Syrian-rebels-of-chemical-weapons-use.html. Accessed Aug 2020
13. BBC News (2015) Islamic State 'used mustard gas' against Peshmerga. www.bbc.com/news/world-middle-east-34471237. Accessed Aug 2020

14. Gillman O (2015) ISIS are making and using chemical weapons in Syria and Iraq says US official as horrific pictures of Kurdish soldiers' injuries caused by mustard gas emerge. Daily Mail. www.dailymail.co.uk/news/article-3230295/ISIS-making-using-chemical-weapons-Syria-Iraq-says-official-horrific-pictures-Kurdish-soldiers-injuries-caused-mustard-gas-emerge.html. Accessed Aug 2020
15. Trager R (2015) Failed terror attack raises alarms about chemical plant security. Chemistry World. www.rsc.org/chemistryworld/2015/07/failed-terrorist-attack-chemical-plant-security. Accessed Aug 2020
16. Bale JM, Ackerman G (2005) Recommendations on the development of methodologies and attributes for assessing terrorist threats of WMD terrorism. Center for Nonproliferation Studies, Monterey Institute of International Studies, p 39
17. Kazi R (2011) The correlation between non-state actors and weapons of mass destruction. Connect Q J 10(4):2
18. CNN News (2020) FBI concludes investigation into 2001 anthrax mailings. CNN News. http://edition.cnn.com/2010/CRIME/02/19/fbi.anthrax.report/. Accessed 19 Feb 2010
19. Stanislawski BH (2005) Transnational organized crime, terrorism, and WMD. In: Blum A, Asal V, Wilkenfeld J (eds) Non-state actors, terrorism, and weapons of mass destruction. International studies review 7(7):158–160. https://doi.org/10.1111/j.1521-9488.2005.479_10.x
20. Bale JM, Ackerman G (2005) Recommendation on the development of methodologies and attributes for assessing terrorist threats of WMD terrorism. Center for nonproliferation studies, Monterey Institute of International Studies, p 35
21. Garamone J (2007) Terrorists using chlorine car bombs to intimidate Iraqis. American Forces Press Service. archive.defense.gov/news/newsarticle.aspx?id=46311. Accessed 6 June
22. The Hague Centre for Strategic Studies (2010). Future Issue: The Future of CBRN. 12(3):7–8
23. United Nation Security Council Resolution 1373 para. 2 (2001) 28 September
24. Szechenyi C (1995) Inside the village of Aum Shinrikyo. Moscow Times. www.themoscowtimes.com/news/article/inside-the-village-of-aum-shinrikyo/340092.html. Accessed 22 April
25. Marlo FH (1999) WMD terrorism and US intelligence collection. Terror Polit Viol 11(3):53–71. https://doi.org/10.1080/09546559908427516
26. RT (2015) 'ISIS Branch' seeking to produce chemical weapons—Iraq and US intel. RT. www.rt.com/news/322726-ISIS-chemical-weapons-intel/. Accessed Aug 2020
27. Summers C (2015) Questions over ricin conspiracy. BBC News. http://news.bbc.co.uk/2/hi/uk_news/4433499.stm
28. Flade F (2018) The June 2018 Cologne ricin plot: a new threshold in Jihadi bio terror. CTC Sentinel 11(7)
29. Bale JM, Ackerman G (2005) Recommendations on the development of methodologies and attributes for assessing terrorist threats of WMD terrorism. Center for Nonproliferation Studies, Monterey Institute of International Studies p 43
30. The Hague Criminal Court (2007) Case 2200050906–2 Strafzaak van Anraat para. 8. deeplink.rechtspraak.nl/uitspraak?id=ECLI:NL:GHSGR:2007:BA4676. Accessed 9 May
31. Anthony I (2014) Exports of dual-use chemicals to Syria: an assessment of European Union export controls. Non-proliferation paper no 35, pp 1–13. https://www.sipri.org/publications/2014/eu-nonproliferation-papers/exports-dual-use-chemicals-syria-assessment-european-union-export-controls
32. Byrnes ME, King DA, Tierno PM (2003) Nuclear, chemical, and biological terrorism. Emergency response and public protection. Lewis Publishers, Boca Raton, London, New York
33. Bale JM, Ackerman G (2005) Recommendation on the development of methodologies and attributes for assessing terrorist threats of WMD terrorism. Center for nonproliferation studies, Monterey Institute of International Studies 10–11
34. Coupland R, Loye D (2009) International assistance for victims of use of nuclear, radiological, biological and chemical weapons: time for a reality check? Int Rev Red Cross 91(874):329–340. https://doi.org/10.1017/S1816383109990105

35. Andre R (2013) CBRN protection: managing the threat of chemical, biological, radioactive and nuclear weapons. Wiley-VCH Verlag & Co., p 4
36. Roy K (2012) Hinduism and the ethics of warfare in South Asia: from antiquity to the present. Cambridge University. https://doi.org/10.1017/CBO9781139084116
37. Formally: The protocol for the prohibition of the use in war of asphyxiating, poisonous or other gases, and of bacteriological methods of warfare
38. Bellamy C (2008) Absolute war: Soviet Russia in the Second World War. Alfred a Knopf Inc., New York
39. www.history.navy.mil/faqs/faq104-4.htm. Accessed Sep 2020
40. Cross G (2017) Dirty war: Rhodesia and chemical biological warfare, 1975–1980. Helion & Company, West Midlands (UK)
41. Tucker JB (1997) Evidence Iraq used chemical weapons during the 1991 Persian Gulf war. The Nonproliferation Review, pp 114–22. www.nonproliferation.org/wp-content/uploads/npr/tucker43.pdf. Accessed Sep 2020
42. joebiden.com/2020/09/02/statement-by-vice-president-biden-on-the-poisoning-of-alexey-navalny/. Accessed Sep 2020
43. According to sources including the New York Times, Skripal continued to work for UK SIS until at least 2016 after his exchange with Russian agents in July 2010. He had been at that point pardoned by then President Medvedev
44. www.gov.uk/government/speeches/pm-statement-on-the-salisbury-investigation-5-september-2018. Accessed Sep 2020
45. Bastings I, Leeuw MW Nieuwenhuizen MS, van Oosterhout T, Henrioud A, Vacca V, Mellado R, Mellin L, Svensson I, Ilchmann K, McLeish C, Perry Robinson J, Phillips T, Lemyre L (2007) ASSRBCVUL–Assessment of the vulnerabilities of modern societies to terrorist acts employing radiological, biological or chemical agents with the view to assist in developing preventive and suppressive crisis management strategies. Final report of ESTO project 502476
46. Economic Consequences of Terrorism (2002) OECD Outlook. 71:117–140
47. US Government Accountability Office (2002) Review of studies of the economic impact of the September 11 2001 terrorist attacks on the World Trade Center. Report GAO-02–700R. www.gao.gov/products/gao-02-700r. Accessed Aug 2020
48. Rogers B, Amlot R, Rubin GJ, Wessely S, Krieger K (2007) Mediating the social and psychological impacts of terrorist attacks: the role of risk perception and risk communication. Int Rev Psychiatry 19(3):279–288. https://doi.org/10.1080/09540260701349373
49. European Security Research and Innovation Forum (2009) Final Report. https://ec.europa.eu/growth/content/security-research-final-esrif-report-0_en
50. van den Brink M, Nieuwenhuizen MS (2011) DECOTESSC1–Demonstration of counterterrorism system-of-systems against CBRNE phase 1. Final Report EU project DECOTESSC1

Biological Threats

J. M. Blatny

1 Introduction

Biological agents are generally defined as microorganisms or toxins, including bioregulators and genetically modified agents (synthetic modified agents), that are able to cause disease, damage or death to humans, animals, plants or materials, and destroy the environment and infrastructures.

Microorganisms are usually invisible to the naked eye. Thus, we neither see, nor feel when we are infected by pathogenic microorganisms. Some biological agents are infectious, others not (Fig. 1). They can be naturally transmitted by air, food, water, direct and indirect contact, as well as by vectors such as mosquitos and fleas. Air, the ambient respirable aerosol that we inhale every day, can contain up to 50% "bioaerosols" depending on the type of environment. Aerosols that contain pathogenic airborne bacteria can represent a serious health threat and have shown to cause infectious disease in humans. Even though the majority of airborne bacteria are generally not harmful to healthy humans, knowledge and awareness about airborne transmission of infectious diseases is crucial for preparedness planning and responding to natural, accidental and deliberate (NAD) events of biological agents. The terms "biological warfare agent" or "biological weapon" are also often associated with these types of agents. With respect to bioterrorism, the term "biological threat agent (BTA)" is usually used, whereas a biological weapon implies a proven and tested delivery system along with the agent.

In general, biological agents are classified according to laboratory biosafety and biodefence considerations. From a biodefence point of view, they are often categorized according to the perceived risk they pose to cause disease, public

J. M. Blatny (✉)
Norwegian Defence Research Establishment (Forsvarets Forskningsinstitutt (FFI)), P.O. Box 25, NO2027 Kjeller, Norway
e-mail: janet-martha.blatny@ffi.no

P.D.E. Biggins and D. Chana (eds.), *CBRNE: Challenges in the 21st Century*, Advanced Sciences and Technologies for Security Applications,
https://doi.org/10.1007/978-3-031-17374-5_3

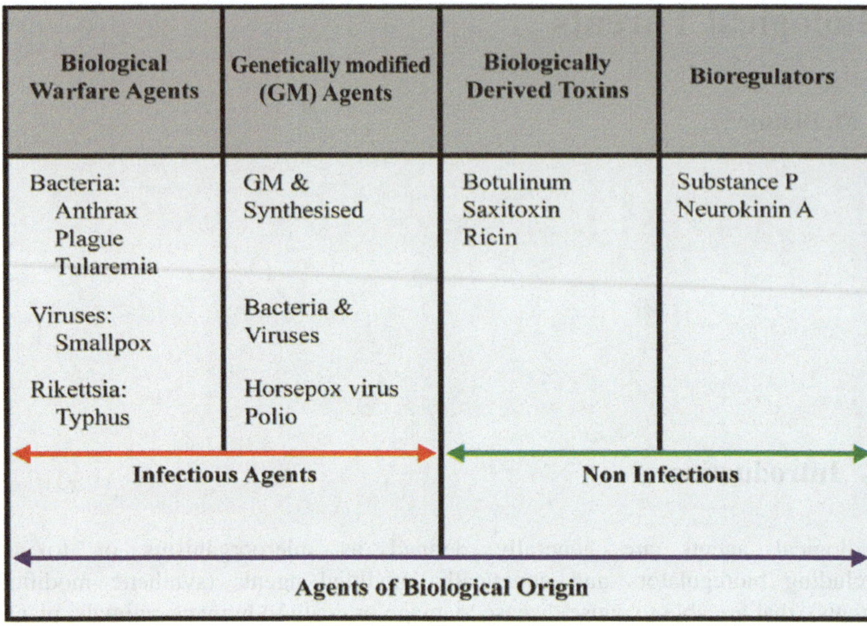

Fig. 1 The Biological Threat Agent Spectrum

health, treatment and national security as a basis for prevention and control mechanisms. The US Centers for Disease Control and Prevention (CDC) classifies biological agents into three categories (A, B and C), where Category A agents are the high-priority agents and pose a risk to national security due to their high human–human transmission ability, in addition to other characteristics of concern: high mortality rate, public health, panic and social disruption. Examples of such agents are those causing anthrax, botulism, plague, smallpox, tularemia, and viral hemorrhagic fever. However, organizations and nations may categorize biological agents and biological threat agents differently from various criteria.

2 Natural, Accidental and Deliberate (NAD) Events

A NAD event with biological agents can occur anytime and anywhere, throughout peace times, during crises and wartimes. In general, biological events are often looked upon as natural events happening during peacetime, and thus setting the baseline for national biological preparedness. A biological natural event can occur simultaneously to additional crisis events and therefore may challenge both national biological preparedness planning and crises management as a whole, e.g. when prioritisation of critical infrastructure resources is required.

Regardless the NAD type of event, prevention and control of biological agents is a national defence and homeland security challenge and must be confronted as such. The COVID-19 pandemic has emphasised the importance of efficient and well-defined national and international prevention and preparedness mechanisms for both public health and critical infrastructure, as well as from a biodefence point of view. Security and safety measures against biological NADs must be a national *investment*, and not regarded as a national cost. This also requires fact-based knowledge and communication to the nation's population in order to gain publics' understanding of the biological risk. The public's trust in the government's ability to protect public health is a critical requirement for achieving an efficient response to biological NAD events. Effective communication from governmental authorities to the population is important for ensuring the population trusts in the government's crises response and for mitigating attempts at fake news and disinformation.

There are several examples where bacteria and viruses were used in biological warfare and deliberate events. Thus, *the risk of future similar events cannot be neglected*. The world is still recovering and learning from the 2001 U.S. anthrax attack, while recovering from the COVID-19 natural event.

2.1 Deliberate

The period between the two World Wars and well into the 1970's was a time when considerable research into the weaponisation of biological agents was carried out at the Nation State level. The bio-weaponeers at that time were aware that air was the most effective transmission route for infecting large populations: by bioaerosols. Thus, several bioweapon programmes focused on constructing and weaponizing airborne microorganisms.

Fact Box: Biological Warfare

Biological warfare has been a tool of war for hundreds of years and the first report is from the 14th century at Caffa stated to cause the beginning of the Black Death, which was the greatest health disaster in our time. During the 20th century, several state-sponsored biological warfare programmes, including in the US and the former Soviet Union, developed and stockpiled massive arsenals of biological weapons. Before the Biological and Toxin Weapons Convention (BTWC) came into force in 1975, the choice of agents in the biological weapons programmes of the US and the USSR was strikingly similar. The "golden age" of biological weapons (BW) stretched from the period between the two world wars and well into the 1970s. Biological weapons were effective as tested population killers with well-documented effects. Agents causing anthrax, plague, smallpox and botulinum toxin were regarded as the most suitable and feared in the BW programmes by the US, the USSR, Germany, Japan, the UK, and Canada regarding pathogenicity and virulence. *Bacillus anthracis* (anthrax) was also first choice when Iraq started developing a BW capability in the late 1980s. Iraq had produced 19,000 litres of botulism toxin, 8,000 litres of anthrax and 2,000 litres of aflatoxins during their BW programme, discovered by the United Nations Special Commission (UNSCOM). Offensive biological weapons capabilities were framed as "first strike" options to destabilize the counterparts and enemy ground forces that would not easily be deterred by conventional ordnance. History has shown that ricin, HIV-infected blood, and food contaminants (for example *Salmonella* species and *Shigella* species) were frequently used for deliberate use. In 2001, the dissemination of anthrax spores by letters and the postal processing and distribution centers in the US resulted in twenty-two cases of anthrax, of which five of the inhalation cases were fatal, and more than 2000 treated with antibiotics in one of the postal facilities [1–4].

Biological weapons are regarded as the only other weapons of mass destruction (WMD) to be able to match nuclear weapons in the scale of casualties, due to the potential for contagion to spread in an uncontrolled manner. A 2000-fold difference can be estimated in the costs associated with civilian casualties when comparing conventional and biological weapons. Therefore, biological weapons are noted as the "poor man's atom bomb."

Biological threat agents (BTA's) are relatively easily acquired as they are naturally occurring. Nation State offensive biological weapons programmes utilising State resources and facilities proved it possible to make biological weapons at scale. Equipment that can be used for the production of biological agents at both smaller and larger scales are so-called dual-use equipment. Examples include equipment used in civilian dairy and pharmaceutical production facilities. Even with dual-use equipment, it can be difficult to develop an effective "biological weapon" without State aid. A "successful" use of biological threat agents to cause infection will also depend on the perpetrator's knowledge, skills and abilities. In 1994–95, the Japanese religious cult Aum Shinrikyo attempted to disperse *B. anthracis* spores in Japanese cities, but despite massive investments, failed due to their unintended use of a non-virulent bacterial strain [5].

Open reports and historical records on far-right Chemical, Biological, Radiological and Nuclear (CBRN) terrorism are scarce, but from 1970 to 2018, 31 incidents from far-right actors were reported [6, 7]. The Norwegian lone-wolf actor Anders Behring Breivik stated his interest in CBRN in his manifest. Actors that successfully use BTAs for intentional purposes may encourage others to do likewise, despite the time needed to perform such actions. Breivik inspired several other terrorists: the New Zealand Mosque attacker (2019) and an American officer planning a larger terror incident in the US (2019). Breivik himself was inspired by the Four Freedoms Community. In 2021, a 16-year-old boy in Norway was sentenced to prison for planning a terrorist act, with potential use of chemical and biological agents and explosives, being affiliated with ISIS. There is no reason to rule out the use of BTAs in deliberate acts. Even failed or crude attacks can have significant consequences and countermeasures must be in place.

These events highlight the consequences for not considering the potential and impact of biological agents for national defence and homeland security. This is an ongoing concern. National budgetary constraints cannot be a limiting factor for improving national security, including knowledge building and biological preparedness. The COVID-19 crisis can, to some extent, be comparable to a hybrid threat situation. Both situations require a conceptual analytical framework to better understand the parameters, mechanisms and dynamics of the multi-domain challenges and implications such events cause for national security and defence.

There are several examples of sole actors not belonging to a terrorist or political group. Bruce Edwards Ivins, a senior biodefense researcher at the United States Army Medical Research Institute of Infectious Diseases (USAMRIID), Fort Detrick, Maryland was alleged to be responsible for the anthrax mail attacks in the US in 2001 and was using institutional resources, according to the US Department of Justice. Sole actors are difficult to identify prior to the incidents, unless intelligence information or information from the public (close family, friends and colleagues) are shared with the police.

In 2018, terrorist plots by jihadists were apprehended in France and Germany where ricin was planned to be used. Investigations of the 2016 Brussel attack revealed that two of the plotters had recorded ten hours of surveillance on the director of the Belgium nuclear programme, with investigators believing they spied with the intent of kidnapping him and entering the nuclear plant where he worked. Ricin is naturally available and produced in seeds of the castor oil plant *Ricinus communis.* It is an extremely highly potent toxin; there are no approved therapeutics treatment. Ricin has been used in suicide attempts, has been disguised and delivered in letters, and used in the 1978 assassination of Bulgarian defector Georgi Markov; a ricin filled pellet was fired into Markov's leg using an umbrella [8]. Ten days before the attack on Markov, the Bulgarian defector Kostov survived a similar attack when standing in an elevator in Paris. In 1981, CIA double agent Korczak was shot in the kidney by an air gun with a pellet containing ricin, also surviving. Ricin has also been used in attempts to poison US politicians by mail, as with the 2001 anthrax attack, demonstrating the urgent need for forensic and identification procedures to be in place. Ricin's use in the treatment of tumors illustrates the dual-use dilemma of scientific work.

Biological deliberate acts may be performed by state-, non-state-, proxy- and single actors any time during peacetimes, crises and wartimes. Current and future concerns are actors who also intend to use genetically engineered modified microorganisms in deliberate events. Synthetic biology technologies provide more possibilities than earlier times and further challenge current response mechanisms.

The biological threat can never be ignored as long as it remains possible to misuse and deliberately disseminate such agents. A joint defence and security approach is required, involving civilian-military collaboration in terms of a whole governmental approach. This would be of benefit for national security, health and biological preparedness. This approach requires also international collaboration and multinational agreements. A whole government comprehensive multi-domain and cross-sectorial approach, including authorities and knowledge builders (e.g. research institutes and academia) and other relevant stakeholders, will as such require a common understanding of the biological threat, risks and consequences. Increasing national biosecurity also increases national biosafety and biological preparedness, as well as supporting non-proliferation agreements internationally.

The risk perception and the threat assessments of biological deliberate events may differ between nations. These assessments usually found the basis for national biological preparedness planning, especially within the biodefence community. Therefore, national biological NAD preparedness planning will differ between nations.

2.2 *Accidental*

The unintentional release of anthrax spores from a biological warfare research facility in Yekaterinburg (formerly known as Sverdlovsk), Russia in 1979 unwittingly demonstrated the effectiveness of infection by inhalation. This was a human failure linked to the maintenance of the facility's ventilation systems [9].

The UK outbreak of foot and mouth disease in 2001 led to the culling of around six million sheep and cows. This can also be regarded as an accidental event, though a pig farm owner was found guilty of starting the outbreak due to not informing the authorities about an incidence of the disease and feeding their animals with untreated waste. The outbreak was not a "typical" deliberate act as it can be assumed that the consequences of the event were not intended.. There was a great impact from this event on national agriculture and tourism, causing economic losses of around 8 billion pounds [10], clearly demonstrating second order effects from a biological event. In 2003, a monkey pox outbreak happened in the USA following the importation of infected animals from Ghana. The source of the outbreak was traced back to native prairie dogs (Cynomys spp.) housed with imported African rodents that infected prairie dogs at an Illinois pet distributor [11]. In 2014, more than forty litres of concentrated live poliovirus solution was accidentally released into the Belgium sewage system by a vaccine producing company [12].

Legionella outbreaks have appeared accidently, spreading from cooling towers and biological wastewater plants, causing fatalities. The outbreaks in Norway 2005 and 2007 clearly demonstrated the benefits of interdisciplinary approaches for source-tracking, detection and identification of an outbreak. It was assumed that an industrial cooling tower caused more than 60 cases and 10 fatalities, but thorough investigations proved that the source was in fact biological wastewater ponds at the industrial site [13]. A similar incident happened in the Bronx, US [14]. Many legionella cases have been reported worldwide, mainly arising from cooling towers, hot tubs/tanks and water fountains.

Germany investigated one of their largest food contaminations events in 2011, which had a tremendous global and European impact [15]. The *Escherichia coli* source strain, *E. coli* O104:H4, was identified in bean sprouts; almost 400 people were infected and there were around 50 deaths. During the search for the contagious agent, cucumbers, lettuce and bean sprouts from Spain, Bavaria and the Netherlands, respectively, were investigated with none identifying the outbreak strain. At the beginning of the analysis, it was stated that the cause of the outbreak was linked to cucumbers from Spain, causing a huge economic impact within Spanish exports: about 200 million US dollars per week. The investigation took several months and impacted food export within the EU, with Russia even banning the import of vegetables from the EU during this period. At least seven European nations were economically impacted due to loss of exports, vegetables rotting in the fields, compensation to the farmers, human loss and absences from the workplace due to sickness. The outbreak clearly demonstrated the second order effects from a biological accidental event. Better surveillance and early warning systems were called for following this serious incident. There was even speculation that the outbreak could have been a deliberate act.

The above examples illustrate that distinguishing between natural, accidental and deliberate events is not always easy. Even the deliberate 1984 Oregon Salmonella outbreak initiated by the Rajneesh Cult was originally noted as an accidental food poisoning event.

Accidental and natural biological events commonly occur and are normal societal challenges. This emphasises the importance of preventive measurements and multi-disciplinary and organizational biological preparedness planning mechanisms towards NAD events.

2.3 Responding to Biological NAD Events

To minimize the effects of a biological NAD event, national authorities need to be aware of the threat, risk and consequences that biological agents pose to society. Physicians must be trained to recognize clinical symptoms resulting from infections caused by biological (threat) agents, as exemplified by the Amerithrax investigation of the U.S. anthrax postal event in 2001. Biological threat and risk assessments focusing on deliberate use of biological threat agents are provided by national authorities, intelligence and police security services.

Informal and formal collaboration between authorities and stakeholders assessing national threats and risks, medical treatment and biodefence, and public health is a pre-requisite in obtaining well-defined reliable biological threat assessments and cost-effective NAD preparedness planning and response mechanisms.

Within the biodefence community, terms such as detection and identification are commonly used. Detection relates to the presence or absence of biological agents, while identification refers to identifying the precise nature of the agent, such as the specific strain of a microorganism. Early detection and rapid identification are both critical for minimisation of casualties and illness, and to affect timely and efficient response, vaccination strategies, and containment to limit transmission of the outbreak. The detection and identification methods for analysing biological agents are similar in use regardless type of NAD event. Genetic sequencing and whole genome sequencing are now common techniques in identification analysis, providing rapid and precise identification of the biological agent. Microbiological techniques provide necessary capabilities to characterize the biological agent that is not obtained by genetic approaches such as viability of the microorganism. Identifying the agent is a main crucial step in obtaining a good situational awareness to set further response actions.

One of the first steps in managing a biological event is to ascertain the NAD feature of the incident: is it a real threat or a hoax, a deliberate, accidental or natural outbreak. When a biological NAD event occurs, extensive investigation is needed: identification of the biological agent and analysis of a wide range of samples from both the environment and the individuals exposed, which may also include microbial forensic analysis. One major challenge is ensuring capabilities to identify an unknown biological (threat) agent in various samples (see Fact Box Identifying Unknowns).

Fact Box: Identifying Unknowns

Identifying unknown CBRE agents in samples

Hoax letters are continuously distributed worldwide. Some of these contain a biological threat agent, while others not. The content of such letters must be identified for several reasons: to confirm the presence or absence of the biological threat agent, to initiate immediately medical treatment for those exposed and to ensure safety measures also for the infrastructure owners. The infrastructure owners need to communicate about the risk and measures taken to their employees such about if the letter contains anthrax, if the building is safe, and the exposure of their fellow colleagues. Hoax letters can contain a mixture of several biological threat agents, or even a mixture of CBR agents and explosives (E). It is often perceived and assumed that the content is anthrax spores. This is not necessarily the case. Several hoax letters have proven to contain ricin. This highlights the need for highly qualified laboratories that are able to analyze the letters and the samples for the presence of a single or a combination of CBRE agents. These analyses require defined protocols, safety measurements, sophisticated expertise and training. Advanced identification methodologies are crucial to enable successfully analysis of samples with an unknown content of CBRE agents. Such identification schemes are also important for verification purposes of relevance to the Biological and Toxin Weapons Convention (BTWC) and Chemical Weapons Convention (CWC). The skill to identify and distinguish between biological NAD events requires also well-defined forensic and attribution procedures as emphasized by the Secretary-General's Mechanism for investigation of alleged use of chemical, bacteriological (biological) or toxin weapons (UNSGM).

2.3.1 An Integrated Warning System

A well-defined biological preparedness system requires efficient and alert medical service systems and emergency support teams addressing the incident *after* it has occurred as well as necessary capabilities to alert authorities prior to such incidents. If the event is deliberate, intelligence, police security services and public health authorities are involved in the investigations. In some cases, even the biodefence community takes part in the investigations.

The strain causing the deliberate Salmonella outbreak in Oregon in 1984 was identified after 4 days, but it took more than one year to verify that this was a biological deliberate event, following the confession of a Rajneesh cult member [16]. Rapid identification methods assist in reducing time to verify the outbreak strain, but additional information is required to prove the type of event.

Biological agents may be difficult to detect before clinical symptoms appear as such symptoms may appear several hours, days or even weeks after exposure. Traditional surveillance systems serve public health well in detecting and responding to infectious disease outbreaks before diagnosis is confirmed or prior to the onset of symptoms. One step in reducing the impact of outbreaks and diseases is to strengthen and develop continent-wide disease surveillance systems. Syndromic-based surveillance systems are in use worldwide daily as nations take part in surveillance of national communicable diseases through, for example, the European Centre for Disease Prevention and Control (ECDC) and the World Health Organization (WHO). The systems are often based on diseases and symptoms that have already occurred and/or where responses have been initiated. An optimal goal is a surveillance system that detects and identifies the biological (threat) agent before humans, animals, or crops are exposed. Systems and control programmes detecting and targeting early-stage outbreaks would result in national cost savings compared to the costs of responding to and recovering from larger outbreaks [17]. Monitoring systems that solely rely on clinical recognition will not prevent exposure to individuals by themselves, as countermeasures would only be implemented once the outbreak has been recognised. The COVID-19 pandemic illustrates this well: restrictions in worldwide travelling were implemented before prophylactic treatments were in place.

Nevertheless, early and rapid identification of the biological (threat) agents before clinical symptoms appear greatly assists in decreasing infection rates, casualties and individuals exposed, allowing adequate medical and protection measures to mitigate the impact of the event [18].

The first attempts at establishing and using early warning biological detection systems were done within the defence communities in order to protect the war fighter from biological airborne attacks during military operations. These systems target aerosols that can contain contagious biological agents, as air is regarded as the most efficient way for dispersion of biological threats. Both UK and US deployed expedient biological detection systems during the first Gulf War. During the 1990's, the US, UK and Canada developed more advanced systems for biological detection.

Several studies have investigated the bioaerosol background in various environments: the presence of bioaerosols in air poses health risks at industrial sites (pollution), rural and urban areas, subway stations and maritime environments [19–21]. The global MetaSUB study increased knowledge about the microbiome profile and their impact on an individuals' health in subway stations in 45 cities worldwide [22]. Several scientific studies have analysed the presence of COVID-19 virus in air, in hospital rooms, subway stations and public areas to gain more fact-finding knowledge about the transmission of the virus. A better understanding of the bioaerosol content in various environments is important for occupational health, public health and biodefence studies, and for the development of biological early warning detection and identification schemes. Furthermore, increased knowledge about aerosol physics and bioaerosol features has proven to be crucial in understanding airborne transmission of COVID-19 and providing recommendations with respect to infectious disease control and prevention: for the use of face coverings, for quality assurance of medical and community face masks and for social distancing.

2.3.2 Modelling and Simulation of Biological Events

It is not easy to prepare for the next transmissible disease, pandemic, or biological NAD event. Epidemiological models are critical planning tools for policymakers, clinicians, and public health practitioners to assist in understanding the consequences of a disease transmitting among the population. These models aid preparedness planning but do not necessarily deal with uncertainties. However, a common feature is the limitation of a model's accuracy due to the state of knowledge of the biological agent, the development and the spread of the transmission (droplet-characteristics), the human immunity, the infective dose and the infection rate. Countermeasures such as medical treatment, vaccination and mitigation strategies will take place during an outbreak. Such measures will have an impact on the development of the disease, and models need to take these into account to provide reliable prediction assessments.

Aerosols that penetrate and deposit in the human alveolar region are of particular concern, but the risk of an individual being infected also depends on the exposure dose of biological agent present in the inhaled aerosols and the individual's susceptibility to being infected. *In other words, modelling is constrained by what we know and what we assume.* The more knowledge gained about a virus' ability to transmit and infect humans, and its survivability under various conditions and environments, the better predictions and guidance we achieve with these models. Scientists still lack sufficient knowledge about aerosol transmission and survival rates of infectious diseases, but several studies on airborne transmission were published during the COVID-19 pandemic. This knowledge is crucial for obtaining efficient, valid and reliable modelling and prediction tools for preparedness planning, environmental and personal exposures.

Infection prevention and control guidelines depend on understanding the route of transmission. The COVID-19 virus transmits by multiple routes: contact, droplet and airborne transmission. Airborne transmission was not initially regarded as a main infection route, and thus was not taken sufficiently into account during risk assessment plans. Infection prevention and controls established during the early phase of the COVID-19 pandemic were based on close human-to-human transmission, contact transmission and experience gained from studying the transmission of the viral respiratory illnesses SARS and MERS (WHO). Some of the evidence for aerosol transmission of COVID-19 is based on epidemiological studies; outbreaks from a Guangzhou restaurant also serving as indirect evidence of aerosol transmission [23, 24].

Airborne transmission of the SARS-CoV-2 virus was a particular knowledge-gap at the start of the COVID-19 pandemic. It was later shown that this virus is present in respired aerosols of less than 5 μm in diameter, that the virus transmits through aerosol particles less than 5 μm in diameter, and that virus-containing aerosols are produced during respiration, vocalization, and coughing [25]. Thorough scientific studies have shown that respiratory droplets smaller than 50 μm can be transported far away from the source. Droplets larger than 150 μm will rarely spread more than two meters upon normal breathing. Evidence is provided that the risk of COVID-19 infection is greatest within two meters distance of an infected source [26, 27]. Droplets between 50–150 μm depend on the air humidity and temperature to provide airborne transmission. Reliable modelling tools (dispersion and epidemiological) for infection prevention, control and biological preparedness planning for such diseases need to include evidence-based data, such as virus-specific biological and physical parameters, transport modelling and human–human infection rate data.

2.3.3 Indoor Early Warning System

Europeans spend more than 80% of their time indoors, but most of the research on dispersion and detection of chemical and biological agents has been performed outdoors. The European Research Network for Critical Infrastructure Protection (ERNCIP) invited a group of experts to review emerging technologies for detection and to assess the detection of airborne chemical and biological agents in critical infrastructures and indoor environments.. The group of experts also considered chemical and biological airborne dispersion modelling tools to implement detectors indoors based on:

(i) various NAD scenarios,
(ii) use of chemical and biological sensors from a cost perspective,
(iii) toxicological or infectious doses,
(iv) mathematical dispersion modelling (Computational fluid dynamics CFD) to locate the sensors at the best place for the chemical and biological sensors in the infrastructures.

Can buildings be efficiently protected by NAD biological releases or incidents, if possible, by physical security and by smart infrastructure design? Karakitsios et al. [28] concluded that sufficient protection against chemical and biological (threat) agents indoors can be obtained by using integrative systems that combine multiple type sensors and dispersion modelling tools. This approach builds on the work of DARPA's Smart building programme in the early 2000's.

During COVID-19, several studies addressed airborne dispersion and transmission of the coronavirus; raising questions about the transmission of the virus through speech and exhaled breath, the stability of the virus in aerosols, the number of viruses per aerosol, the dispersion distance of the virus, and the impact of the ventilation flow on airborne transmission in restaurants [29]. Researchers from Kobe University and the research organisation Riken used the world's fastest supercomputer Fugaku, to model how the coronavirus spreads in a room when people are sitting at a dining table and speaking without face masks. The COVID-19 crisis has highlighted the use of such modelling and simulation computer tools for protective and biological preparedness purposes, advocating for the inclusion of these tools in interdisciplinary approaches.

Current technologies allow for miniaturization of sensors that can further be integrated into materials, building materials and wallpaper: a multifunctional wallpaper for automatic alarm upon fire disasters has proven helpful for public safety by saving human lives [30]. New material science could, in the future, integrate biosensors into wallpapers to detect airborne infectious agents. It is possible to imagine biosensors in wallpapers connected to emergency operation centers as early warning systems. Seismic wallpapers are available where data registers movements, which is for strengthening civilian infrastructure vulnerable to earthquakes. Electronic textiles for smart fabrics are becoming reality. These textiles incorporate electronics and computational elements into the fabric material, which can improve diagnostic and early medical treatments [31].

Integrating (bio)sensors into clothing is already recognized as "smart wearable technologies." Current wearable technologies [32] can monitor a wide range of health parameters: heart rate, body temperature, respiration and glucose levels. Based on these parameters, algorithms are being developed to determine chemical, biological and radiological exposures. First responders' and warfighters' can benefit from integrating chemical, biological and radiological sensors into their clothing. Exposure to chemical, biological and radiological agents can be monitored and the agent can rapidly be identified in order to take preventive and follow up measures.

In future smart cities where the Internet of Things (IoT) is the hub for digital living, it is reasonable to foresee that future biological detection and identification schemes will also be a part of this loop. This is a measure to protect society, infrastructure and the population.

3 Pandemics During the 20th and 21st Centuries

Human-to-human disease transmission causing endemics and pandemics is one of society's major concerns. Pandemics are biological natural events. This is clearly demonstrated through examples of several flu virus transmissions through history and by the spread of smallpox. The World Health Organization (WHO) officially acknowledged smallpox as eradicated in 1980, considered to be the twentieth century's biggest achievement in international public health. The availability (or lack thereof) of smallpox stocks (see fact box Smallpox) and anti-vaccination movements towards the smallpox vaccination programme raise concerns for future outbreaks and societal ability to rapidly respond to this highly transmissible disease. Anti-vaccination movements have been ongoing for hundreds of years. In more modern times, such movements have been targeted towards vaccination programmes for diphtheria, tetanus, and pertussis (DTP) and measles, mumps, and rubella (MMR). Use of social media has facilitated anti-vaccination supporters against COVID-19 to spread misinformation rapidly about vaccine safety causing concern about reduction in vaccination rates.

Fact Box: Smallpox

Smallpox Two stocks of smallpox for vaccine purpose are secured at the Center for Disease Control and Prevention (CDC) in the US and the State Research Center of Virology and Biotechnology (VECTOR) in Russia. There are debates whether the stockpiles are to be retained or destroyed. WHO established an Advisory Group of Independent Experts (AGIES) to provide advice on matters of strategy and evaluation of the research programme, and to assess whether additional research with live variola virus is necessary from the public health perspective [33]. In 2013, AGIES concluded that there was no reason to retain live variola virus for the further development of smallpox vaccines, antiviral agents and for further development of diagnosis. In 2019, the decision about destroying the stocks were reaffirmed by the WHO, but the timing was not decided [34]. As synthetic biology technologies continue to develop, it makes it possible to create variola virus using publicly available information and common laboratory procedures. Earlier investigations and reviews by an Institute of Medicine (IOM) committee concluded in 1999 the need for live smallpox (variola) virus to develop medical countermeasures, vaccines, therapeutics, and diagnostic tools in case of accidental or deliberate releases. They further recommended that CDC and VECTOR perform complete genomic sequencing of the retained smallpox virus isolates to gain knowledge of sequence patterns and variations. In the late 1990s both the scientific and public health communities had found both scientific and civil defence reasons for retaining the stocks of live smallpox virus. A lesson learned from the Iraq war in 2003, was the need for information sharing among public health authorities, the public domain and military intelligence as the lack of this caused reasons for suspicions and concern about the smallpox vaccine programme. Why reintroducing the vaccine in the absence of the disease? In 2002, President George W Bush announced reviving smallpox vaccination programmes, for military and (voluntary) civilians. Smallpox is still categorized as a biological warfare and biological threat agent.

When reviewing the period between 1940 and 2004, most of the emerging or re-emerging infectious diseases appeared after the 1970s [35]. Since then major societal changes have occurred on a global level. The rapid human to human transmission of COVID-19 is primarily due to international travel, social non-distancing and lack of host immunity to the virus. The fight against such a rapid contagious disease is challenging, but nations worldwide focused on obtaining COVID-19 vaccines as fast as possible. Within a rather short time, the world was able to produce several COVID-19 vaccines. However, the vaccine programmes highlighted inequalities across the globe in terms of vaccine production and supply. Another example of rapid vaccine development is the production of vaccines against Ebola. During the Ebola outbreak, the Bill & Melinda Gates Foundation supported funding to scale up emergency efforts to rapidly obtain an Ebola vaccine that were put into trial phases in 2013–2016. If there is a strong societal need, political will and ability, success will be achieved.

Several European nations have succeeded in their initial societal prevention strategies and response to the COVID-19 pandemic to keep infection rates as low as possible and not to strain their health care systems. The success has also relied on and benefitted from the populations' trust in the government, the follow-up of governmental guidelines and restrictions, the availability of digital technology to rapidly communicate and disperse important information, and good health care systems in many nations.

Despite the rapid transmission of COVID-19, which has caused more than 491 million people to be infected and 6,155,344 deaths as of the 5th of April 2022 [36], COVID-19 is not the worst Flu-like pandemic the world has seen. The 1918 Spanish Flu resulted in 40–50 million deaths. The 2009/2010 Swine Flu resulted in almost 600 000 deaths, comparable to annual deaths from seasonal flu at a global level. Some studies estimate that actual infected Swine Flu cases numbered 700 million −1.2 billion, but following the Swine Flu pandemic, laboratory confirmed cases were estimated at around 18500 globally (WHO). These estimates illustrate the complexity in providing accurate numbers of confirmed cases and deaths as results from the disease as there are many factors that must be considered during such calculations: symptoms, underlying illnesses, misattribution of cause of deaths, laboratory confirmation and the test identification analysis itself.

The severe acute respiratory syndrome (SARS) was initially reported in February 2003 in Asia. SARS is a viral respiratory illness in humans caused by the coronavirus SARS-CoV, also a virus that had crossed the species barrier from animals to humans [37]. Globally, approximately 800 deaths were recorded amongst 26 countries within months of the first emergence of the virus, but still the disease was controllable from an epidemiological point of view. In comparison, the SARS-CoV-2 virus causing COVID-19 has a higher infection rate than SARS-CoV causing SARS [38].

Bacteria have also caused pandemics, most famously *Yersinia pestis*, which caused the Black Death and led to 75–200 million deaths globally. The bacteria *Mycobacterium tuberculosis* causing tuberculosis is among the top ten death causes in the world resulting in 1,5 million deaths annually. Tuberculosis is an infectious disease and the highest death tolls of tuberculosis are within the African region. Among new tuberculosis cases in 2014, about 3% were estimated as being multidrug-resistant (WHO). The increase in antibiotic-resistant bacteria is worrying with respect to human global health and food security, and so-called 'superbugs' are increasingly widespread. Antimicrobial resistance may prove to be one of our greatest threats if we do not continue to combat antibiotic-resistant bacteria and develop new medical treatments.

Travelling and social non-distancing are key factors contributing to the increase of COVID-19 transmission. During the Spanish Flu period, travelling was limited, but not social contact. Knowledge about virus and disease transmission, hygiene routines and rapid communication mechanisms were at that time limited. Fortunately, twenty-first century data technologies, digital communication methods, and social media have allowed rapid communication and information sharing about protection measurements to the population. One of the worst scenarios our modern world has been waiting for has happened (COVID-19), however, it has also highlighted the second order effects on national and international economies and trade in the highly interconnected and interdependent globalized world.

Many virus-causing diseases in humans originate from animals (zoonosis). Bats are the natural host of the Ebola and Corona virus. Other examples of zoonotic diseases include Anthrax, Lyme, Rabies, West Nile virus, Avian Influenza viruses, and the mosquito-borne viral infections Dengue and Zika. According to the Center for Disease Control and Prevention (CDC), it is estimated that three in four new or emerging infectious diseases in humans are transmitted from animals. Zoonotic diseases kill more than two million people per year. Most human infections with animal influenza arise from close, direct contact with poultry or swine. This argues for global surveillance of animal diseases as well as human diseases as zoonosis is a public health problem (One Health Approach, WHO). Twenty-first century pandemics have raised awareness about livestock-human close contact in societies, but animal husbandry has a long tradition and dates back several thousand years. Modern animal husbandry focuses on production systems and selective breeding as well as wildlife farming for food and trade. Modern animal husbandry may lead to a reduction of livestock genetic diversity and therefore a decrease in disease resistance amongst livestock. Deforestation may also impact virus emergence as tropical forest edges are a major platform for novel human viruses [17].

The largest outbreaks of Ebola hemorrhagic fever appeared in 2014–2016 in West Africa, and smaller outbreaks still occur. Lessons learned from the Ebola epidemic related to the lack of international and local preparedness, the importance of understanding local customs, medical infrastructures and knowledge about disease transmission. Somewhat similar to COVID-19. The Dengue and Zika diseases are regarded as travel-associated infections. It is debatable if we are sufficiently able to implement effectively lessons learned about countermeasures from previous endemics to combat new endemic and pandemics.

3.1 Growing Population Impacts Transmission

The Swine Flu (2009/2010) primarily affected children, young adults and those under 60 years of age. The age distribution of COVID-19 cases has changed during the pandemic, initially primarily affecting older adults to primarily affecting younger adults. This can be due to many factors such as identifying cases with severe symptoms and/or mild cases, advances in testing capacities and mutations of the virus. It is difficult to predict which groups and individuals a new Flu–like virus will target. This challenges preparedness planning. The world's population is growing older, with persons over age 65 being the fastest-growing age group [39]. The UN World Population Prospects report 2019 predicts that by 2050, one out of four persons in Europe and North America will be 65 years in age or over, while those over 80 years in age will have tripled from 143 to 426 million.

The 2019 UN report estimates a 26% increase in the global population by 2050 (from 7.7 billion to 9.7 billion). However, growth rates vary greatly among regions. India is expected to overtake China as the most globally populous country within a decade. The majority of growing populations are in the developing countries and where health care systems are not well-developed and will thus suffer in responding to a pandemic. Furthermore, the growing population will cause an increase in urbanization. By 2050, more than 60% of the world's population is expected to live in cities. The dense population in urban environments will thus led to increased transmission of diseases. New approaches to risk management and preparedness planning are needed, potentially in "smart cities" (see Sect. 7.1). The trend towards "future smart cities" using information communication technology, and IoT built into the fabric of the city infrastructure has the potential to assist in rapid response to a pandemic, allowing, self-containment, quarantine and isolation and provision of healthcare in a smart city environment.

4 Biological Preparedness and Technology

Technological developments in a wide spectrum of fields would benefit current and future biological preparedness. The main developing areas are within sensor and digital technologies, particularly information and communication technologies (ICT), but also materials technology.

4.1 Future Bio-sensor Technology

Future biological detection and identification systems with improved algorithms used in an orthogonal architecture is assumed to provide lower false alarm rates with a higher level of confidence. The challenge of these systems is to distinguish between naturally occurring microorganisms and a deliberate release of biological agents. To aid sensor technology, design systems need to characterise the environment where they are deployed. This additional information will provide ambient flux details, which can alert the system to anomalies in the respirable air of the deployed location. This is useful for supporting decision-making processes with regard to whether a deliberate event has occurred and whether to provide necessary countermeasures. The cost of such systems needs to be considered when deploying; if it also provides a secondary public health benefit then it is an investment worth making. Such systems will produce a huge amount of data, which can be used to characterise and categorise airborne environments, and provide baselines.

We have entered the "Big Data era," where new techniques to deal with the volume, variety and visualization of gathered data are being developed, and that are highly supportive to knowledge and information management processes. With data connectivity, the growth of social media (see Sect. 7.2 with respect to influence) and IoT, machine learning algorithms now offer new possibilities of how to process and use data. This will enable artificial intelligence (AI) systems along with "Big Data" to contribute to the design of sensors and therapeutics. AI approaches can enable sensing platforms to collect and process information obtained by surveillance/monitoring systems in a more robust way than currently possible, and thereby improving biological detection and identification systems.

The miniaturisation of sensor technology is opening up a new era. The advent of wrist band health state trackers that enable monitoring individuals' own physiology during fitness and sports demonstrates the concept's potential, however, it does raise societal issues about privacy and governance of personal private data.

Such sensors can be further developed with new sensing materials and electronics and provide new analytical capabilities. New materials for developing biosensors are already available, such as carbon tubes, polymers, bioelectronics, and along with AI allows the development for targeted smart biosensors and improve monitoring and biological situational awareness. Developments within quantum technologies and biology will also push forward better sensor schemes and smaller biosensor devices. New sensors will improve detection and identification capabilities of biological agents and biomarkers. Biosensors will at an individual level be used for real-time awareness of "the human state" for over-training, nutritional deficiencies, immunocompetence, cardiac health and muscular-skeletal injury and CBRN exposure.

Increased use of biosensors will enhance human capabilities. Such changes could include ocular enhancements to eyesight through implants or contact lenses. For instance, expanding human vision past the visible spectrum of light. Neuroscientists have suggested that neural enhancements of the human brain for data transfer is possible and where direct communications between humans and machines, and with other humans, could optimize command and control systems [40]. This is often called the "cyborg-human." Such disruptive technology will probably be available by 2050. The applications would first of all be valuable for military and first responders, but does require a debate about legal, security, and ethical frameworks before taken into common use.

Human microchipping, in which a circuit device containing information or transmitting signals is implanted, is already in use. The first experiments were done in 1998, but in 2009 the British scientist Gasson proved how a human microchip could be hacked. The Swedish incubator Epicenter began microchipping its employees in 2015 to let them operate printers and other office equipment. A radio frequency identification chip opens a door with just a wave of your hand in front of a chip reader. Human chips can function as contactless credit cards, key cards, and rail cards making (urban) life easier (SmartCity). The UK firm BioTeq offered with a low cost (£70 and £260 per person, 2018) the implants to businesses and individuals and has fitted 150 implants in the UK to open their doors, access their office or start their car with a wave of their hand, and store medical data. These applications have raised an alarm in the UK that firms are taking more control over their employees and interfering in their privacy. Again: who and what controls the human being? These are examples of a chip, or a device implanted in the body for non-medical purposes (in contrast to a pacemaker, as an example). It is possible that future human microchips designed to identify exposures to diseases and CBRN as an early warning detection system will develop. This makes the human being become a monitored mobile sensor. It raises considerable concerns in terms of biohacking and in general hacking of the chip, just as computer hackers hack computers and biohackers hack anything biological.

Legal and ethical aspects relating to the individual, group, firm and the nation are now raised more than ever in relation to the technological advances discussed above, such as access to personal information and the right to use such information in various settings (e.g., insurance). The strategic drivers for future developments and security at both an individual and national level are no longer at the biological nor digital level, they are linked to law enforcement, ethics and human behaviour.

Additive manufacturing and the development of sensors allow buildings and infrastructures to be 3D printed with integrated biosensors or health monitoring sensors. 3D printing is already in use to produce organs: 3D bioprinting. So far, bioprinting has demonstrated limited success and especially within the area of skin and bone, but the technology is relevant to regenerative medicine to address the need for tissues and 3D printed organs for transplant (e.g., kidney and liver).

Other type of sensors are airport scanners. Airport scanners monitoring fever prevalence have been tested and, in some cases, implemented in several Asian nations. These scanners are to detect febrile passengers, as fever is one noticeable sign of Flu-like infections, but various other factors such as age and outdoor temperatures may affect the detection on febrile arrivals and provide false alarms [41, 42]. The challenge, and the need, is still to verify the cause of the fever and to rule out or confirm that the diagnosis is a result of a biological NAD event.

Personalized medicine is in its advent in terms of tailored medical treatment, therapy and preventive care. It addresses how a person's unique molecular and genetic profile make them susceptible to certain diseases and how they would respond to specific treatments. Future medicine will utilise more proactive treatment regimens; reducing costs and enhancing quality of life enhancement, with each individual receiving their full genomic information as a part of their individual medical record, allowing physicians and clinicians to implement tailored healthcare approaches [43]. Advanced genetics and proteomics will be used to establish tools for the early detection of CBRN exposure and biomarkers: some are currently used for early detection of cancer. The aspiration is to identify illness caused by "all" biological agents before symptoms appear.

4.2 Digital Approaches for Biological Preparedness

The storage of personal data about our lives are used for customized advertisements through internet and social media. The information can be sold or given to other parties that want to obtain commercial benefits by such targeted advertisements. In the above section, human microchips, sensors for data collection and storage and the cyborg human were mentioned, all providing sensitive and private information. Authorities can take advantage of the availability and accessibility of private information. China has implemented a social credit system to monitor the behavior of its population and ranks them based on their "social credit". The more information the authority can access, the more unwanted control from the individual's freedom and democracy perspective is obtained. Smart speakers in the home could be potentially hacked to perform a biological release if a perpetrator has control over your digitally connected devices. Digital hacking is concern at a national security level in terms of protection of the critical national infrastructures. For instance, what if water cleaning plants are attacked? The cleansing process of drinking water plants are often secured and controlled by digital systems. The well-known Stuxnet attack on Iran's uranium enrichment facility, attributed to the US and Israel, has shown that it is possible to manipulate systems controlling a nuclear reactor. There is no reason a similar approach cannot occur with biological agents resulting in disease transmission.

5 Preparedness

The COVID-19 response also benefitted from lessons learned from the Swine Flu, Avian Flu and the SARS outbreaks (2002–2004), all biological natural events. It seems that we can be sufficiently prepared against new infectious diseases. This struggle derives from a lack of available knowledge at the time of the event, including e.g., facts about the genetic feature of the new virus, transmission and infection rate, and available and rapid production of effective vaccines.

A common approach is often to prepare for "yesterday's crises" as these are "the known events" and to assess short term biological preparedness planning. The ability to acknowledge and address "black swans" and uncertainties in a changing world is a "must" when dealing with crises preparedness. This also requires knowledge and research to gain scientific evidence to support the planning. A state's preparedness ability to respond to, or at least plan for, unexpected and surprising events is also a measure of the nation's ability of awareness and capability to act upon preparing for biological natural events and pandemics. COVID-19 showed that we can no longer prepare for consequences from what are considered as "normal" events. What is "normal"? Are new emerging diseases a "new normal"? A new normal might be a society that accepts more "unknown unknowns" and unexpected events. The Global Risks Report 2020 [44] emphasizes that "turbulence is the new normal". Politicians and governmental decision-makers will benefit from a population who trusts their authorities and who are able to face unexpected events. Authorities need to be able to communicate the risks, the truth and consequences of scenarios, situations and crises to their population, so everyone is prepared and understands possible impacts on health, society, economical consequence, and national security. We need to learn how to prepare for longer crises events which last months.

Though history has proven that infectious diseases frequently appear as natural events, COVID-19, Spanish, Avian- and Swine Flu, and SARS can also be regarded as so-called "low probability—high consequence" events. It has been well-known, and prior to the COVID-19 outbreak, that a pandemic would happen but not the timing of the event. This can to some extent be compared to a CBRN event, which is commonly referred to as a "low probable-high consequent event." CBRN events are regarded as horrific, and the negative impacts so challenging that their risks and consequences may be too difficult to address, and are in the worst case neglected. The COVID-19 pandemic illustrates some aspects of a deliberate large-scale CBRN event. However, economically, the costs in prevention are significantly lower compared to the costs of responding to and recovering from such events. Therefore, such scenarios are too costly to be neglected, both at a national and global level.

Using scenarios to address biological preparedness and planning is a common tool among stakeholders and authorities, especially within the defence communities. Event 201 was one such scenario, but it did not anticipate shortfalls of world leaders that rejected advice from public health authorities [45]. Biosecurity experts often use military-like exercises for biological NAD preparedness. The military war-gaming approach may assist in addressing public health scenarios as the Defence communities and the Armed forces have a long tradition in preparing towards a major crisis. From that perspective, the civilian authorities can learn from military approaches and tools for preparedness planning. Studies show that the huge economic losses of COVID-19 are not necessarily just associated with a high health impact. In many countries where there is a large loss in national industrial production, the COVID-19 health impact is low in terms of deaths [46].

The impact of pandemics on national and the global economies can vary as it depends on additional factors, such as concurrent ongoing crises. During the Swine Flu outbreak, a global economic crisis was already ongoing. Both Swine Flu and COVID-19 caused closure of schools and negatively impacted global tourism. The increased use and development of digital tools aided schools and universities in maintaining teaching and lecturing students at home. However, there were negative effects of school closure. For instance, children forced to stay home with abusive family members. Worldwide losses of approximately 5 trillion dollars were estimated for 2020, highlighting the need for further research on the costs of crises and the return of investments in preventive strategies [17].

The complex interdependencies in our modern societies must be considered when addressing risk assessments, governance and impacts of major disruptions in our nations, both at a group and an individual level. This requires social-technical understanding and a systems thinking approach. In 2019, OECD published a paper on systems approaches in policymaking [47] and promoting cross-sectorial and multidisciplinary collaboration; the collaboration needed between "hard" science and social science and methodology to manage systems behaviour, risk and complexity as to better understand the non-linear behaviour of complex systems. A social-technical system analysis approach is part of the twenty-first century paradigm to address the consequences of the changing landscape.

6 Biosecurity, International Control Mechanisms and Convergence

Over the past two decades, synthetic biology has emerged as a powerful approach for studying and manipulating biological systems; combining cellular biology and genetic engineering techniques with the principles of systems engineering. Synthetic biology is described as "designing and constructing biological modules, biological systems, and biological machines or, re-design of existing biological systems for useful purposes."

Synthetic biology provides both threats and opportunities, challenging existing conventions and biosecurity frameworks. The ability to manipulate biological systems with ever greater precision naturally raises the dual-use risk. The possibility to create new biological threats agents and biological weapons that are more contagious and transmissible than currently existing is a risk. A concern is that manipulated biological agents will not respond to treatment with existing antibiotics or anti-viral drugs, will escape from current detection and identification schemes, and their environmental persistence and infection rate will increase. Synthetic biology causes concern due to its dual-use potential and misuse by terrorists.

The use of synthetic biology to engineer and manipulate biology can play a key role in developing new countermeasures, new identification and detection schemes, medical treatment, and vaccines. Future possibilities include combining biological engineering, synthetic biology, genetic design and AI to create biological materials with engineered properties, such as living biosensors.

Concern can also be raised about novel synthetically modified microorganisms which can be either deliberately or by accident released into the environment and disrupt the normal biodiversity (see fact box The Convention on Biological Diversity). This would have devastating effects, and negatively impact public health, the economy and national security.

Fact Box: The Convention on Biological Diversity

The Convention on Biological Diversity (CBD)

CBD was signed in 1992 and focuses on sustainability and societal security; people's need for food security, medicines, fresh air and water, shelter, and a clean and healthy environment. The background for this convention is the potential and the emerging loss of biodiversity in nature, thereby reducing nature's goods and services for humans (www.cbd.int).

There are many examples of synthesised or genetically manipulated biological agents (Fact box Synthesized or manipulated biological agents). DNA fragments need to be synthesised in order to (re)create microorganisms, and the purchase of such synthetic genes from commercial institutions is commonly performed within scientific communities. Unfortunately, not all commercial institutions synthesising DNA fragments cross-reference ordered sequences for similarities to sequences found in pathogenic microorganisms. Companies supplying synthetic DNA fragments are encouraged to check and compare ordered nucleotide sequences for similar sequences to those found in pathogenic microorganisms as an important step in reducing the threat of (re)creating pathogenic biological agents. The International Gene Synthesis Consortium (IGSC) (DNA synthesis companies) have devoted resources to screen customers and their requests for potential security issues, even though no legal mandate requires them to do so. In January 2020, the Nuclear Threat Initiative (NTI) and the World Economic Forum recommended establishing a common DNA sequence screening mechanism, as a follow up from IGSC. It is in the best interests of the companies that they invest in biosecurity aspects.

Fact Box: Synthesised or Manipulated Biological Agents

Examples of synthesized or manipulated biological agents

Several viruses have been synthesized, including the polio virus in 2002, the H1N1 1918 Spanish Flu in 2005, the West Nile virus in 2010, and SARS in 2008. Scientists have genetically altered the H5N1 avian influenza genome, resulting in an infective virus capable of airborne passage between ferrets. After passaging the genetically modified H5N1 virus between ferrets a highly pathogenic virus was obtained and with the potential to be transmissible between humans [48]. This raised great concerns about the consequences if the virus were to escape the laboratory or fall into the hands of bioterrorists.

The horsepox virus was synthetically constructed by chemicals (2018) where DNA fragments were purchased from a company that produce chemical synthesized DNA. Viruses have smaller genomes than bacteria (<30 kb), it has been demonstrated that new cells of *Mycoplasma mycoides* may be created in vitro (1,08Mb genome) and thereby changing the species of the bacterium [50].

In 2018, the first genome-edited baby was produced using the CRISPR technology on embryos [51]. This experiment raised the need for regulations and efforts among the scientific communities using and controlling CRISPR technology. Human embryo editing was pure science fiction in the 1980's and is now a reality.

Synthetic biology challenges the biosecurity framework and therefore current policies need to be revised. As a part of students' training, awareness about biosecurity, national and international conventions (Fact box international control mechanisms and conventions) should now be included. The technology needed for genetic engineering is available for purchase on the second-hand market. Synthetic genes and genetically modified cells could be constructed in a hobbyist's basement or garage, so-called DIY biology (do-it-yourself biology): biological threat agents could be designed and (re)created by grass-roots biohackers, not just by highly skilled scientists.

Fact Box: International Control Mechanisms and Conventions

International Control Mechanisms and Conventions

The Convention on the Prohibition of the Development, Production and Stockpiling of Bacteriological (Biological) and Toxin Weapons and on their Destruction (the Biological and Toxin Weapons Convention BTWC) was the first multilateral disarmament treaty banning the production of an entire category of weapons (https://www.un.org/disarmament/biological-weapons/). The BTWC was established as a supplement to the 1925 Geneva Protocol which prohibits use, but not possession or development of chemical and biological weapons (https://www.un.org/disarmament/wmd/bio/1925-geneva-protocol/).

The Geneva Protocol was to ban weapons of war: the use of asphyxiating, poisonous, or other gases, as these substances had been used during World War 1 causing close to one million casualties. BTWC entered into force 26 March 1975 but has no verification regime and thus does not allow the intrusive inspections as the Chemical Weapons Convention (CWC) permits. Despite conventions banning the production and use of biological agents, speculations about biological warfare state programmes are ongoing and the possibility cannot be neglected.

Comprehensive national safeguards have been put in place in many countries to prevent misuse of biological threat agents, these are often recognized as arms control agreements. One example is the "Australia Group (AG)" (www.australiagroup.net), which through harmonization of export controls, seeks to ensure that exports from participating countries do not contribute to the development of chemical or biological weapons. AG is not a formal decisive group and its actions are limited to non-proliferation measures. Another example is the multilateral export control regime "The Wassenaar Arrangement on Export Controls for Conventional Arms and Dual-Use Goods and Technologies. This group focuses on transfers of conventional arms and dual-use goods and technologies, and sensitive technology/materials (e.g., stealth). But the most important and overall agreement is the 2004 "UN Security Council Resolution 1540" (UNSCR 1540), which covers "proliferation of nuclear, chemical and biological weapons, as well as their means of delivery, that constitutes a threat to international peace and security", also for terrorist purposes. The UNSCR 1540 requires all States to adopt and enforce laws to obtain this effect and other measures to prevent the proliferation of CBRN weapons and their means of delivery to non-State actors (www.un.org).

6.1 Convergence of the BTWC/CWC

Synthetic biology challenges the convergence between the biological and chemical threat spectrum. As biological agents can be synthesized by chemicals and chemicals may be produced by using biosynthetic approaches and advanced biotechnological methods, these issues have an impact on the chemicals and biological agents listed by CWC and BTWC (Fact box International Control Mechanisms and Conventions), respectively. Both conventions include toxins and other 'mid-spectrum' agents (Fig 3.1). While CWC refers specifically to the listed toxins ricin and saxitoxin regarding verification purposes, ricin is listed on both conventions: BTWC and CWC.

However, categorisation of synthetic biological agents is not clear. The Organisation for the Prohibition of Chemical Weapons (OPCW) shows concern about how metabolic engineering and synthetic biology can be misused to synthesize toxins, toxic chemicals or precursors in quantities that could pose a threat. This highlights the convergence between BTWC and CWC. From such a perspective, one can argue that these two Conventions could converge to a certain extent as categorising synthetic agents may not be clear cut due to their dual- and/or multiple-origin and scaling-up production technologies.

As mentioned in the beginning of this chapter, biological agents are generally categorized according to laboratory biosafety and biodefence measures, as well as from an export control perspective (Australia Group). However, there are inconsistencies between the various lists and categories provided by WHO, EU, CDC, National Institutes of Health (NIH) and different nations, even though the CDC list is the most commonly recognized and widely used list among scientific, biodefence and health communities [52]. Thus, a need for an internationally agreed list of categories of biological agents from both a biosafety and biodefence perspective are welcomed.

7 Changing Landscape, Threat Space and the Grey Zone

The changing landscape represents a set of various challenges and developments: old and new threats and actors; traditional and emerging diseases; emerging disruptive technologies; social-cultural changes; security political changes at both national and international levels; and world political leaders representing new or altered national ambitions.

Nations are more interdependent than ever due to international trade and economies. Nations are no longer self-sufficient with livestock, food, technology and other necessary goods and products for their population to be safe and secured. Commercial products such as vaccines are developed and produced in selected nations, and trade and international agreements between nations is necessary. The second order impacts of COVID-19 clearly illustrated such dependencies: nations' access to medical and protective equipment became a challenge as well as quality assurance mechanisms of personal protective products according to national requirements.

Societal and national security are becoming more complex and vulnerable to trade, technology and ownership, which furthermore impacts national critical national infrastructure. Digital products and solutions, produced by a nation, are not necessarily trusted by other nations due to lack of information control and access to data. Data, and Big data, is the current and new golden age. The need to protect and secure biological data and genetic information is highlighted.

The five nations with the largest defence budgets (US, Russia, China, India, Saudi Arabia) were unprepared for and hit hard by COVID-19. The changing landscape includes also addressing budgets combatting pandemics and climate changes, as addressed by the UN Sustainable Development Goals and Paris Agreement 2015. Increasing defence budgets does not implicitly raise national security against pandemics.

Non-proliferation initiatives are being challenged as Nations withdraw from agreements causing global insecurity. The US stated their withdrawal from the 1987 Intermediate-Range Nuclear Forces (INF) Treaty in 2019 and announced their withdrawal from WHO in 2020 [53]. Many nations worldwide have not ratified the BTWC. The possibilities that synthetic biology opens for the development of new products and agents, and the presence of naturally occurring biological agents that cause pandemics or rapidly cause fatalities and illness is regarded with interest by various actors and terrorists. The BTWC still does not have a verification protocol, biosecurity and biosafety measures are not necessarily in place worldwide, there are challenges in identifying a biological NAD event, synthetic biology challenges current identification schemes, and there is a lack of appropriate medical treatment: the deliberate use of biological agents therefore remains a concern. This concern encompasses both threats and risks from state and non-state actors and their intention to execute such events. As stated by the sixteenth century writer John Lyly: "all is fair in love and war".

Verification regimes must involve socio-technological approaches. Successful verification is based on reliable information provided by those (nations and individuals) taking part in the verification and the technology used. Thus, verifiable treaties contain a web of constraints and provisions designed to deter cheating and to explore challenges of trust as part of verification regimes due to the balance between (i) truth and confidence and (ii) human-technology and social-cultural layers.

Emerging disruptive technologies challenge even further the trust between humans and technology. This also illustrates the need for knowledge about human behaviour within biological treaty regimes.

COVID-19 has highlighted a broader set of threats and challenges than the world was ready for, even though planning exercises in various Nations had been held prior to the event. We were not able to "address all the unknowns" and predict shutdowns of modern societies, increasing use of home schools, rising unemployment rates, large impacts on tourism and trade, and the rapid need to implement new laws. COVID-19 is regarded as the largest peacetime crisis from a population security and safety and national economic perspective. Asymmetrical crises and scenarios, especially in the grey zone between the boundaries of security and defence, will create higher uncertainty.

The COVID-19 pandemic has caused immense socio-economic interruptions, health challenges and global security political challenges. We no longer know what the "normal situation" is as new and old actors change continuously in the current landscape. The new normal might be looking at a society's ability to deal with "new daily surprises" in an era of disrupting developing technology.

Two areas are further elaborated below as they represent two major drivers for our future: smart cities and social media platforms. Future biological events and preparedness must be seen in conjunction with these drivers.

7.1 Smart Cities

Smart cities use information communication technology and IoT to improve efficiency, sustainability and citizen welfare. Such cities also aim to become cleaner, reduce traffic and improve urban life. Modern societies are more sensitive to unexpected events, with a high media exposure, reflecting the emerging role of social networks. Social media platforms are important tools in unfolding civil crisis and disaster response. The platforms become a critical capability for information flow from authorities, rumours, emerging activities and events from adversarial actors. The use of social media can both escalate and de-escalate a (biological) crisis depending on the information provided. Whether we like it or not, social media and digital communication platforms are a normal part of our lives for both business and pleasure from a biological threat perspective; the easy and simple use of uncritically releasing and sharing information about a NAD event will result in dramatic consequences and cause fear among the population. Crowds of people could be entering hospitals for medical treatment and assistance, illustrating the necessity of triage schemes at all hospitals. On the other hand, social media platforms assist in providing rapid relevant information and necessary communication from authorities and first responders to the population for efficient protective measurements. Government and private agencies, as well as private persons, can purchase influencers to use social media for their PR stunts on YouTube, Facebook etc., and increase "likes" on Facebook etc. These approaches can also be a result of an aggressive adversarial information campaign. As for any event, digital communication platforms must be used with caution in crises. Awareness towards adversary actors on social media platforms must be a part of our daily safety and security measures.

The society's critical national infrastructure is vulnerable to incidents or attacks against digital communication schemes. This requires that these vulnerabilities must be part of the planning process when addressing risk management, preparedness planning and assessing biological threats. Disinformation can be spread by social media and can be used for crowd manipulation and generating social hysteria. The understanding of how information technology and how to navigate the information environment, is of vital interest not only from a biological threat perspective but also from a wider threat space.

A Smart city taking advantage of technological improvements and data sharing can promote facilitating vehicle parking, efficient energy consumption, improving traffic flow, smarter security, improving waste management, and disaster planning. But the crucial point is the connection of all of these fragmented technological platforms in a "systems approach" to address vulnerabilities for preparedness planning. Building a Smart city out of an existing city would require renewing and building completely new infrastructures. In near future (1–20 years) it is likely that Smart cities would resemble today's cities but managed more efficiently. At an emergency event, buildings could communicate to provide crucial information to first responders and the population, thus mitigating the consequence of a biological NAD event depending on integrated sensors that are able to identify the biological agent and even clinical symptoms.

7.2 *Information and Influence*

Social media is a digital platform for spreading information, misinformation and disinformation. Disinformation refers to false or manipulated information, which is knowingly shared to cause harm in contrast to misinformation, which refers to false or manipulated information spread without the knowledge that it is false. Social media, crossing geographical and demographic boundaries, has given actors in inter- or intra-state conflicts new channels to try to affect opinions worldwide [54]. Social media uses the benefit of the human and cognitive aspect linked to the cyber domain (cyber-social); the so-called socio-technical perspective and where individuals are being influenced by actors to achieve different tasks by manipulating social media users. The COVID-19 pandemic is the first major global health event in which digital platforms are the major distributor of information. This pandemic showed that decision makers had to deal with this crisis without having specialist knowledge relating to dis- and misinformation.

Authorities have had serious concern about disinformation on COVID-19 as it has an impact on citizens' freedom, public security and safety. Social media platforms may have fake owners. The technology that started out to connect people worldwide has also resulted in platforms with mistrust causing fear and anxiety in our societies. This can further cause mistrust to authorities during biological NAD events. Our democratic future relies on open societies to distribute reliable information through social media technology ensuring our safety and security, but also including control mechanisms avoiding actors providing disinformation. This cyber domain and digital technology will continue to grow, and further measures include prevention strategies, fast detection and response mechanisms to hinder unexpected and unwanted events disrupting public safety and security, regardless of any kind of biological NAD event.

8 Summary

The future of bioterrorism and deliberate use of biological agents is extremely difficult to predict. Similarly, for natural and accidental biological events. Bioterrorism is a part of the global terrorism posing a threat to world security. Uncertainties are linked to potential ongoing state-sponsored biological (warfare) programmes since monitoring of clandestine offensive efforts is complicated by the dual-use nature of equipment, technology and research of concern. Adversary actors, terrorists and terrorist organizations may be even harder to deter and, as they are not bound by the BTWC treaty, it can be assumed that terrorists have fewer constraints on the use of biological (threat) agents, though their technical skills and capabilities are likely to be variable. The skills among terrorists seem to increase and terrorist organizations seem to possess trained microbiologists and chemists to answer the call to arms. Lessons learned is that whatever technology is available will eventually be applied in deliberate events or warfare as one side seeks to gain an advantage over the enemy and the counterpart (i.e. asymmetric conflicts, hybrid).

As organized crime is financed, as various actors control more modern infrastructure in nations, and as knowledge levels increases among the population due to information available and educational structures in society, it cannot be ruled out that future terrorists and other actors with a biological or CBRN interest, capability and intent will increase.

As the American author Walter Russell Mead stated about COVID-19, "the virus has provided the world with an extraordinary demonstration of the power of weaponized biology".

He argued that the twentieth century was the Age of Physics, with the nuclear bomb as the primary weapon, while the twenty-first century "looks now to be an Age of Biology, when the capacity to unleash gene-engineered plagues on one's opponents—or their crops—can provide countries with strategic advantage". During the 2020 UN Counter-terrorism week the UN Secretary-General António Guterres highlighted how the COVID-19 pandemic also effects vulnerabilities to "new and emerging forms of terrorism", as many nations continue to face terrorism threats during the COVID-19 pandemic. He stressed that (quote) "like the virus, terrorism does not respect national borders" and (quote) "non-state actors must not be allowed to exploit the "fissures and fragilities" of rising psycho-social, economic and political stresses, related to the coronavirus." Even in 2005, the US Senator Bill Frist stated at the World Economic Forum, "the greatest existential threat we have in the world today is biological".

Biological events are a part of our everyday life. It is the extent, the intent and the scale of the NAD event that challenges our ability to mitigate, manage and respond to the NAD event sufficiently and timely. Therefore, continuous knowledge building as a crucial step in being prepared for the next NAD event is an evident need. Improved detection and identification schemes and use of new technological approaches will add value to prevent and respond to biological NAD events. Lessons learned from history have shown that "a whole government" approach is needed to combat future biological NAD events. Furthermore, the global changing landscape calls out for a re-thinking in preparedness as well as a social-technical system approach to address the consequences of interconnected crises, including biological.

References

1. Al-Agamy HM (2011) Tools of biological warfare. Res J Microbiol 6(3):193–245. https://doi.org/10.3923/jm.2011.193.245
2. Richardt A, Sabath F (2013) A glance back-myths and facts about CBRN incidents. In: Richardt A, Hűlseweh B, Niemeyer B, Sabath F (eds) CBRN protection: managing the threat of chemical. Biological, Radioactive and Nuclear Weapons. Wiley-VCH Verlag, pp 1–38
3. Riedel S (2004) Biological warfare and bioterrorism: a historical review. BUMC Proceed 17:400–406. https://doi.org/10.1080/08998280.2004.11928002
4. Wheelis M, Rózsa L, Dando M (eds) (2009) Deadly cultures: biological weapons since 1945. Harvard University Press
5. Danzig R, Sageman M, Leighton T, Hough L, Yuki H, Kotani R, Hosford ZM (2012) Aum shinrikyo insights into how terrorists develop biological and chemical weapons. Center New Am Soc, Washington DC, p 31
6. Koehler D, Popella P (2018) Mapping far-right chemical, biological, radiological, and nuclear (CBRN) terrorism efforts in the west: characteristics of plots and perpetrators for future threat assessment. Terrorism Polit Violence 32(8):1–25. https://doi.org/10.1080/09546553.2018.1500365
7. Nesser P, Stenersen A, Oftedal E (2016) Jihadi terrorism in Europe: the IS-effect. Perspect Terrorism 10(6):3–24
8. Moshiri M, Etemad L, Balali-Mood M (2014) A ricin review. Toxinology 1–15
9. Meselson M et al (1994) The Sverdlovsk anthrax outbreak of 1979. Science 266(5188):1202–1208
10. Mclaws M, Ribble C (2007) Description of recent foot and mouth disease outbreaks in nonendemic areas: exploring the relationship between early detection and epidemic size. Can Vet J 48:1051–1062
11. Sklenovská N, Van Ranst M (2018) Emergence of monkeypox as the most important orthopoxvirus infection in humans. Front Public Health 6:241. https://doi.org/10.3389/fpubh.2018.00241
12. Duizer E, Rutjes S, de Roda Husman AM, Schijven J (2016) Risk assessment, risk management and risk-based monitoring following a reported accidental release of poliovirus in Belgium, September to November 2014. Euro Surveill 21(11):30169. https://doi.org/10.2807/1560-7917.ES.2016.21.11.30169
13. Blatny JM, Reif BAP, Skogan, G, Andreassen Ø, Høiby EA, Ask E, Waagen V, Aanonsen D, Aaberge IS and Caugant DA (2008) Tracking Airborne *Legionella* spp. and *Legionella pneumophila* at a biological treatment plant. Environ Sci Technol 42(19):7360–7367
14. Lapierre P, Nazarian E, Zhu Y, Wroblewski D, Saylors A, Passaretti T et al (2015) (2017) Legionnaires' disease outbreak caused by endemic strain of *Legionella pneumophila*, New York, New York, USA. Emerg Infect Dis 23(11):1784–1791. https://doi.org/10.3201/eid2311.170308
15. Mellmann A, Harmsen D, Cummings CA, Zentz EB, Leopold SR, Rico A, Prior K, Szczepanowski R, Ji Y, Zhang W, McLaughlin SF, Henkhaus JK, Leopold B, Karch H (2011) Prospective genomic characterization of the german enterohemorrhagic *Escherichia coli* O104:H4 outbreak by rapid next generation sequencing technology. PLoS ONE 6(7):e22751. https://doi.org/10.1371/journal.pone.0022751
16. Salerno RM et al (2004) A BW risk assessment: historical and technical perspectives. Nonproliferation Rev 11(3):25–55. https://doi.org/10.1080/10736700408436977
17. Dobson AP, Pimm SL, Hannah L, Kaufman L, Ahumada JA, Ando AW, Bernstein A, Busch J, Daszak P, Engelman J, Kinnaird MF, Li BV, Loch-Temzelides T, Lovejoy T, Nowak K, Roehrdanz PR, Vale MM (2020) Ecology and economics for pandemic prevention. Scien369 (6502):379–381. https://doi.org/10.1126/science.abc3189
18. Blatny JM (2007) Detecting and responding to bioterrorism. NATO security through science series a: chemistry and biology. In: Green, MS (ed) Risk assessment and risk communication strategies in bioterrorism preparedness. pp 77–9243

19. Dybwad M, Granum PE, Bruheim P, Blatny JM (2012) Characterization of airborne bacteria at an underground subway station. Appl Environ Microbiol 78:1917–1929. https://doi.org/10.1128/AEM.07212-11
20. Dybwad M, Skogan G, Blatny JM (2013) Temporal variability of the bioaerosol background at a subway station: concentration level, size distribution and diversity of airborne bacteria. Appl Environ Microbiol 80(1):257–270. https://doi.org/10.1128/AEM.02849-13
21. Leung MHY, Tong X, Bøifot KO et al (2021) Characterization of the public transit air microbiome and resistome reveals geographical specificity. Microbiome 9(112):1–19
22. Danko D, Bezdan D, Evan E et al (2021) The International MetaSUB consortium. A global metagenomic map of urban microbiomes and antimicrobial resistance. Cell 184:1–18. https://doi.org/10.1016/j.cell.2021.05.002
23. Wang CC, Prather KA, Sznitman J, Jimenez JL, Lakdawala SS, Tufekci Z, Marr LC (2021) Airborne transmission of respiratory viruses. Science 373(981):1–12. https://doi.org/10.1126/science.abd9149
24. Lu J, Yang Z (2020) COVID-19 outbreak associated with air conditioning in restaurant, Guangzhou, China, 2020. Emerg Infect Dis 26(11):2791–2793. https://doi.org/10.3201/eid2607.200764
25. Lu J, Gu J, Li K, Xu C, Su W, Lai Z et al (2020) COVID-19 outbreak associated with air conditioning in restaurant, Guangzhou, China. Emerg Infect Dis 26(7):1628–1631. https://doi.org/10.3201/eid2607.200764
26. Santarpia JL, Herrera VL, Rivera DN, Ratnesar-Shumate S, Reid P St, Denton PW, Martens JWS, Fang Y, Conoan N, Callahan MV, Lawler JV, Brett-Major DM, Lowe JJ (2020) The infectious nature of patient-generated SARS-CoV-2 Aerosol. https://doi.org/10.1101/2020.07.13.2004163
27. CDC (2020) Scientific brief: SARS-CoV-2 transmission. https://www.cdc.gov/coronavirus/2019-ncov/more/scientific-brief-sars-cov-2.html
28. Karakitsios S, Busker R, Tjärnhage T et al (2020) Challenges on detection, identification and monitoring of indoor airborne chemical-biological agents. Saf Sci 129:1–17. https://doi.org/10.1016/j.ssci.2020.104789
29. National Academies of Sciences, Engineering, and Medicine (2020) Airborne transmission of SARS-CoV-2: Proceedings of a workshop in brief. The National Academies Press, Washington, DC. https://doi.org/10.17226/25958
30. Li Y, Qian H, Hang J, Chen X, Cheng P, Ling H, Wang S, Liang P, Jiansen L, Xiao S, Wei J, Cowling B, Kang M (2021) Probable airborne transmission of SARS-CoV-2 in a poorly ventilated restaurant. Build Environ 196:107788. https://doi.org/10.1016/j.buildenv.2021.107788
31. Chen F, Zhu Y, Dong L, Yang R, Xiong Z (2018) Fire alarm wallpaper based on fire-resistant hydroxyapatite nanowire inorganic paper and graphene oxide thermosensitive sensor Fei-Fei. ACS Nano 12:3159–3171. https://doi.org/10.1021/acsnano.8b00047
32. Dias T (2015) Electronic textiles, 1st edn. Woodhead Publishing, April, Smart Fabrics and Wearable Technology, p 2015
33. WHO (2013) Advisory group of independent experts to review the smallpox research programme (AGIES). Report to the World Health Organization Geneva, Switzerland November 2013
34. WHO (2019) Smallpox eradication: destruction of variola virus stocks. Report by the Director-General. Seventy-Second World Health Assembly A72/28. April 2019
35. Saunders-Hastings PR, Krewski D (2016) Review reviewing the history of pandemic influenza: understanding patterns of emergence and transmission. Pathogens 5(66):1–19. https://doi.org/10.3390/pathogens5040066
36. WHO (2021) World Health Organization: Coronavirus (COVID-19) Dashboard. https://covid19.who.int/
37. Peiris JSM, Phil D, Yuen KY, Osterhaus ADME, Stöhr K (2003) The severe acute respiratory syndrome. N Engl J Med 349:2431–2441. https://doi.org/10.1056/NEJMra032498

38. Zhang L, Zhu J, Wang X, Yang J, Liu XF, Xu X (2021) Characterizing COVID-19 transmission: incubation period, reproduction rate, and multiple-generation spreading. Front Phys 8:588. https://doi.org/10.3389/fphy.2020.589963
39. UN report (2019) World population prospects. https://population.un.org/wpp/
40. NATO Science & Technology Trends 2020–2040. Exploring the S&T Edge NATO Science & Technology Organization. Office of the Chief Scientist, March 2020, www.sto.nato.int
41. Priest PC, Duncan AR, Jennings LC, Baker MG (2011) Thermal image scanning for influenza border screening: results of an airport screening study. PLoS ONE 6(1):e14490. https://doi.org/10.1371/journal.pone.0014490
42. Cho KS, Yoon J (2014) Fever screening and detection of febrile arrivals at an international airport in Korea: association among self-reported fever, infrared thermal camera scanning, and tympanic temperature. Epidemiol Health. https://doi.org/10.4178/epih/e2014004
43. Mathur S, Sutton J (2017) Personalized medicine could transform healthcare. Biomed Rep 7(1):3–5. https://doi.org/10.3892/br.2017.922
44. The Global Risks Report (2020) World economic forum. Insight Report .15th edn. http://www3.weforum.org/docs/WEF_Global_Risk_Report_2020.pdf
45. Maxmen A (2020). Two decades of pandemic war games failed to account for Donald Trump. Nature news features 6 Aug 2020. https://www.scientificamerican.com/article/two-decades-of-pandemic-war-games-failed-to-account-for-donald-trump1/
46. UNIDO (2020) United Nations Industrial Development Organization. https://www.unido.org/stories/coronavirus-economic-impact-10-july-2020
47. OECD. Organisation for Economic Co-operation and Development (2019) Systemic thinking for policy making –the potential of systems analysis for addressing global policy challenges in the 21st century. SG/NAEC(2019)4
48. Herfst S, Schrauwen EJA, Linster M, Chutinimitkul S, de Wit E, Munster VJ, Sorrell EM, Bestebroer TM, Burke DF, Smith DJ, Rimmelzwaan GF, Osterhaus ADME, Fouchier ROM (201) Airborne transmission of influenza A/H5N1 virus between ferrets. Science 336 (6088):1534–1541. https://doi.org/10.1126/science.1213362
49. Noyce RS, Lederman S, Evans DH (2018) Construction of an infectious horsepox virus vaccine from chemically synthesized DNA fragments. PLoS ONE 13(1):e0188453. https://doi.org/10.1371/journal.pone.0188453
50. Gibson DG, Glass JI, Lartigue C, Noskov VN, Chuang R, Algire MA, Benders GA, Montague MG, Ma L, Moodie MA, Merryman C, Vashee S, Krishnakumar R, Assad-Garcia N, Andrews-Pfannkoch C, Denisova EA, Young L, Qi Z, Segall-Shapiro TH, Calvey CH, Parmar PP, Hutchison CA III, Smith HO, Venter JC (2010) Creation of a bacterial cell controlled by a chemically synthesized genome. Science 329(5987):52–56. https://doi.org/10.1126/science.1190719
51. Cloer D (2019) Interview: genetically modified babies: an insider's view. www.vision.org. https://www.vision.org/interview-william-b-hurlbut-genetically-modified-babies-insiders-view-8975
52. Tian D, Zheng T (2014) Comparison and analysis of biological agent category lists based on biosafety and biodefense. PLoS ONE 9(6):e101163. https://doi.org/10.1371/journal.pone.0101163
53. CRS report (2020) U.S. Withdrawal from the World Health Organization: Process and Implications. Congressional Research Service. https://crsreports.congress.govR46575
54. Bergh A (2020) Understanding influence operations in social media: a cyber kill chain approach. J Inf Warfare 19(4):110–131

Radiological and Nuclear

M. Lavelle

Abstract Mapping a radiological or nuclear security risk at the necessary granularity requires a 'live' understanding of six contributing factors: (1) material science; (2) engineering processes; (3) transport; (4) storage; (5) human factors and (6) cyber-physical controls. Operating at this level of specialist knowledge will need a shared human—machine approach, likely taking the form of a global radiological and nuclear information management system built on modern graph processes.

1 Introduction

Explosives, along with some chemical and biological weapons, can all be made in a domestic kitchen. In contrast, radiological and nuclear (RN) materials[1] emitting energies high enough to significantly impact human health and the environment are the product of either: (1) a nuclear reactor; (2) a particle accelerator or (3) on an industrial scale separation and enrichment process (Fig 1). This alone should make RN materials a relatively low proliferation risk. Well, not quite.

[1] The exact definition of radiological and nuclear materials can vary widely, depending on use case. For our purposes, we define a radiological material as containing any isotope of an element with an unstable atomic nucleus that is capable of spontaneously emitting energy via radioactive decay (see 4.1.1). Nuclear materials form a subset of this group, made up of isotopes capable of sustaining a nuclear fission chain reaction. Technically, this means isotopes of the 15 actinide elements with an odd neutron number. In practice, we discuss only uranium-235, U-233 and plutonium-239 since they are frequently used to generate an explosive fission chain reaction within a nuclear weapon. Overall, our focus is on higher energy ionizing radiation sources forming an acute security risk, rather than the chronic health and environmental impacts of naturally occurring radiation (e.g. radon).

M. Lavelle (✉)
Intelligent Ontologies Ltd. & Institute for Security Science and Technology, Imperial College, London SW7 2AZ, UK
e-mail: m.lavelle@imperial.ac.uk

P.D.E. Biggins and D. Chana (eds.), *CBRNE: Challenges in the 21st Century*, Advanced Sciences and Technologies for Security Applications,
https://doi.org/10.1007/978-3-031-17374-5_4

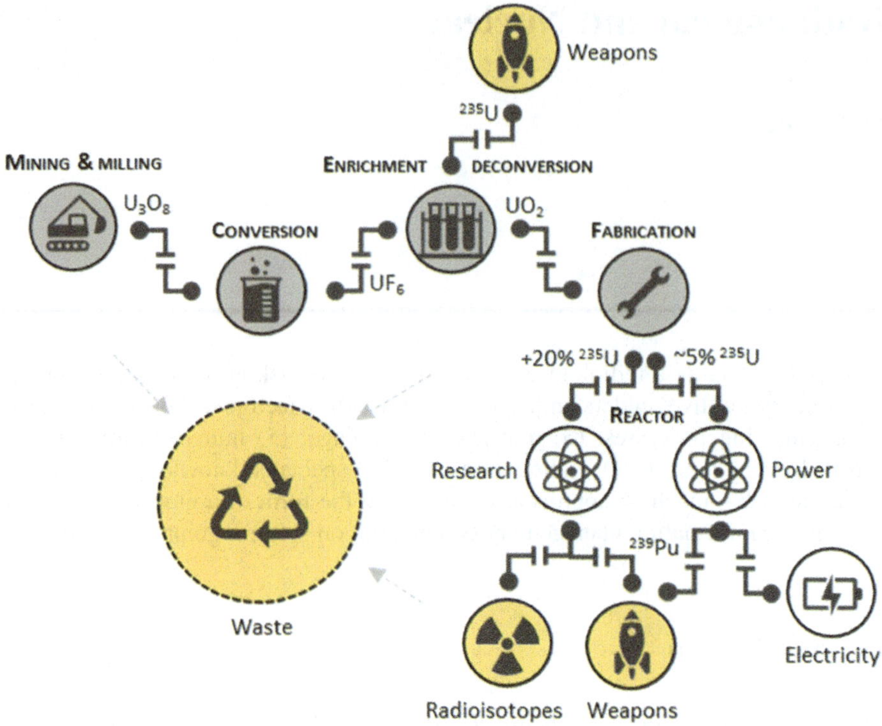

Fig. 1 Production pathways for radiological and nuclear materials of prime security concern (yellow shading)

While there is zero commercial demand for sarin or anthrax, the global nuclear medicine radioisotope[2] market was approximately $17 billion in 2020 [1]. Over 50-million nuclear medicine procedures take place each year, across ~10,000 accredited hospitals [2]. Add the use of radioisotopes in industrial tracers, gauges and inspection, plus the supply of some 18 million enriched uranium fuel pellets[3] to the world's + 400 nuclear power reactors every year, and the risk is clear: the world is awash with radiological and nuclear materials.

---

[2] In common usage the term is interchangeable with radionuclide and radioactive isotope. Radioisotope refers to any isotope of an element whose atoms contain an unbalanced, excess amount of energy. Isotopes are chemical variants of the same element, each with different numbers of neutrons. For example, strontium has four stable naturally occurring isotopes (Sr-84, 86, 87 & 88) and thirty-one unstable radioisotopes spanning Sr-73 to Sr-107. Strontium-89 (used in cancer treatments) and Sr-90 (thermoelectric power generators) are produced during fission and are of security concern.

[3] For a single, standard pressurised water reactor generating 1GWe/yr. Globally, this requires the processing of ~ 70,000 tonnes of natural uranium oxide every year (https://world-nuclear.org/information-library/facts-and-figures/world-nuclear-power-reactors-and-uranium-requireme.aspx July 2021 update).

1.1 Ionising Radiation

Nuclear reactors and particle accelerators generate five commonly used fission isotopes and ~40 activation-derived radioisotopes.[4] All share one key attribute: high-energy, unstable atomic nuclei. Their commercial use, and public danger, comes from the shedding of this excess energy as the isotope decays to a lower energy, stable form. The energy is emitted via any combination of the following:

Alpha decay: if the instability is a result of the nucleus having too much mass, it ejects two protons and two neutrons in the form of a helium nucleus. Common alpha emitters include isotopes of uranium and plutonium (used in power generation and nuclear weapons), americium (smoke detectors) and polonium (anti-static devices and satellite thermoelectrics).

Beta decay: if the nucleus has an unbalanced ratio of neutrons to protons, it can either (i) create and eject an electron and an antineutrino, transforming a problematic neutron into a proton; or (ii) create and eject a positron along with a neutrino to convert a proton to a neutron. Common beta emitters include isotopes of strontium (thermoelectric power generators and industrial inspection), nickel (piezoelectric power generators), phosphorous (DNA tracers) and fluorine (PET imagining).

Gamma decay: after undergoing an initial alpha or beta decay, some nuclei will continue to dissipate their remaining excess energy by emitting gamma rays—massless short wavelength photons from the highest energy limits of the electromagnetic spectrum. Common gamma emitters include isotopes of cobalt and caesium (industrial sensors), barium (food preservation—initiated by beta decay from caesium) and technetium (medical tracers).[5]

Neutron decay: rarer than the other processes, 'free' neutrons can be ejected from the nucleus during nuclear fission/fusion or in certain particle accelerators. Free neutrons are unstable, decaying to a proton, electron and anti-electron-neutrino within 15-min. Controlled neutron imaging is used widely across many industries.

As radioactive elements are constantly changing, knowledge of the entire decay chain is needed to fully understand the material risks. For example, in a strontium-90 thermoelectric generator, much of the radiation emitted (2.28 meV) is derived from the intermediate yttrium-90 decay product, not the original Sr-90 (Fig. 2).

The danger of radiation is primarily from subsequent induced ionization damage in other materials. Where an expelled energetic particle[6] or photon interacts with

[4] Radioactive isotopes of security concern are artificially generated for medical and industrial use via (i) separation of standard reactor fission products; (ii) the controlled neutron bombardment of inserted reactor 'targets'; or (iii) via particle acceleration—often in compact cyclotrons.

[5] Where the unstable excess energy is produced in the electron shell—not the nucleus—X-ray photons are generated instead of gamma. Although x-rays are ionising, they are mostly machine generated and so not discussed in detail here.

[6] In addition to alpha and beta particles, free neutrons also cause ionization, but only indirectly. A stable nucleus will naturally absorb the extra neutron, creating an unstable nucleus. As before, this new excess energy is then shed via standard alpha, beta or gamma decay.

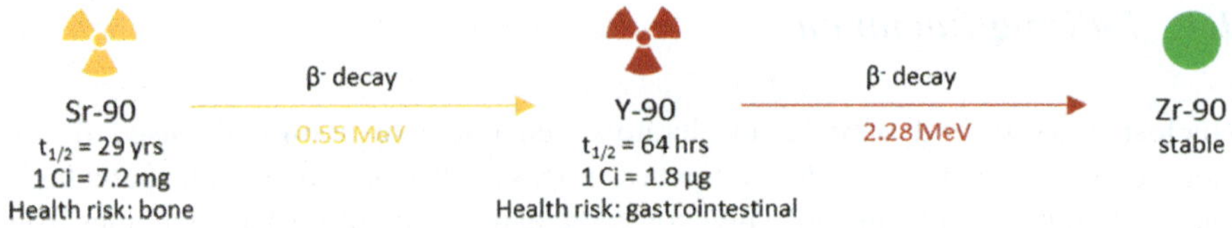

Fig. 2 Understanding material radiation risk. For a Sr-90 source, the main energy emission is linked to the Y-90 to Zr-90 β-decay, not the initial decay of Sr-90. For comparison, 1 Curie (non-SI unit) of radioactivity is contained in a Y-90 mass 1/4000th that of Sr-90

a stable atom, it may remove electrons from the shell, resulting in an unbalanced charge. From the human, animal and plant health perspective, it is this ionization effect that can impact a living cell's ability to replicate accurately—or at high doses may destroy the cell outright. The radiated energy may also poison the cell through the splitting of H and O bonds to form hydrogen peroxide and other highly reactive hydroxyl radicals in cell water. The outcomes range from acute radiation syndrome or an increased long-term cancer risk in humans and animals, to supressed plant growth.

Add the broader issue of environmental radioisotope contamination impacting food chains and water supplies, or radiation damage to critical national infrastructure, and it is clear why these materials must be proactively regulated, constantly monitored and secured from any malicious or accidental loss of control. Unfortunately, history repeatedly shows many current regulatory and monitoring systems are imperfect.

1.2 Safety, Security and Regulation

There are no universally applicable international safety or exemption[7] limits for radioactive materials. Different countries are free to establish their own discharge/dose/exposure limits, although these are usually closely aligned with guidelines issued by the *International Commission on Radiological Protection.* Recognising that human health and safety is the overriding concern, national exposure limits are based on the potential for ionising radiation to cause negative health effects under specific scenarios. These are driven by both the physical properties of a material or device, and partly by the way in which it is used.

In the civil nuclear arena, research and power reactors have been involved in seven serious nuclear accidents impacting human health and the environment: Fukushima (2011), Chernobyl (1986), Three Mile Island (1978), Enrico Fermi Unit 1 (1966), SL-1 (1961), Sodium Reactor Experiment (1959), and Windscale (1957). Although the incidents are few, seven over 54-years, the impacts were significant.

[7] Defined as levels of risk below which regulation is either 'out of scope' (not necessary) or 'light touch' (exempt). For instance, the radiological impact assessments carried out by the International Atomic Energy Agency (IAEA) helps national regulators define such materials by considering a wide variety of possible exposure routes, including water and food pathways.

Fukushima Daiichi Nuclear Disaster (2011)

Following the magnitude 9.0 Great East Japan Earthquake and tsunami, a loss of on site and external power at the coastal Fukushima Daiichi nuclear plant resulted in severe damage to three of the six reactor cores. A linked series of hydrogen explosions damaged three further buildings, releasing significant quantities of radioactive material into the atmosphere and hampered immediate response efforts. Some 120,000 residents within ~40-km of the plant required rehousing. The reactor breaches released iodine-131, cesium-134 and cesium-137 into the environment. Around 20-million-cubic-metres of contaminated soil is now stored as radioactive waste. Following complex filtering to reduce the level of the most active isotopes, 1.2 million litres of contaminated water are expected to be released into the Pacific from holding tanks over the next decade.

To date, Fukushima and Chernobyl are the only Level 7 (maximum) events on the *International Nuclear Event Scale*; Fukushima produced only ~10% of the (I-131 equivalent) radiation released at Chernobyl.

In 2019, thirty-six countries reported a total of 189 cases of nuclear and radioactive material 'outside regulatory control' to the International Atomic Energy Agency's (IAEA) *Incident and Trafficking Database.*[8] This included six incidents of clear trafficking for malicious use, with the remainder classed as having insufficient information to determine intent. The open source NTI database classified fifty incidents in 2019 as theft; forty have yet to be recovered [3].

Since 1993, 3,686 RN incidents have been reported to the IAEA, of which 290 involved a confirmed or likely act of trafficking or malicious use. Twelve of those incidents involved highly enriched uranium and two included plutonium [4].

Samut Prakan Incident (2000)

A cobalt-60 teletherapy source manufactured in Germany in 1969 was installed at a Thai hospital in 1981. The device had an initial activity of 196-TBq (5,300 Ci). The hospital retired the source in 1994 and sold it to the local representative of a Canadian radioisotope supplier. The company stored the device in an unlicensed warehouse, before moving it to an unused carpark near Bangkok. In January 2000, the source was acquired by local scrap dealers and dismantled—by which time the activity of the source had decayed to ~16-TBq.

[8] The IAEA attempts to monitor all types of fissionable nuclear material (uranium, plutonium and thorium), naturally occurring and artificially produced radioisotopes and radioactive contaminated material, such as scrap metal. Countries are also encouraged to report incidents involving scams or hoaxes where material is claimed to be nuclear or otherwise radioactive. See https://www.iaea.org/resources/databases/itdb. The live database is not publicly available.

Over the next 2 weeks, ten people were admitted to hospital with acute radiation syndrome. More than 1,800 people living within 100-m of the scrapyard were exposed to varying levels of excess gamma radiation. Three of the original hospital admissions died. [This followed a very similar 1987 hospital/scrapyard incident in Goiania, Brazil, involving 1,375 Ci of Cs-137].

In 2019, over 80% of the publicly reported incidents were from six countries: Belgium, Canada, France, Japan, South Korea and the USA. Rather than implying unusually poor security, they are amongst the few countries requiring the open reporting of RN-related events.

1.3 Uses and Concerns

Figure 3 highlights the IAEA's top three (of five) categories of common radiological device classified as 'dangerous to the person'.

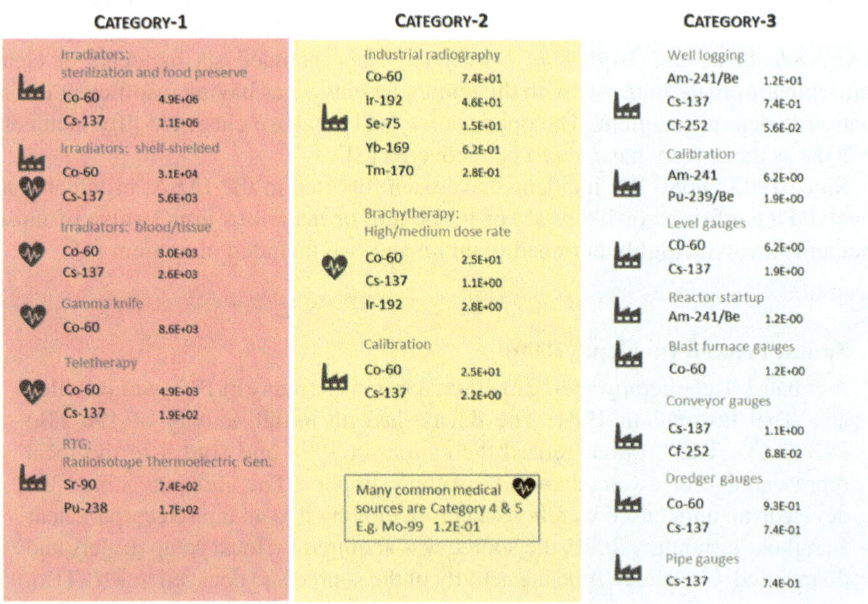

Fig. 3 IAEA categories of manufactured radioactive sources of greatest security concern. The 00E + 00 values are typical activity ratio (A/D) for devices in common use. Icons highlight industrial vs medical uses. Adapted from International Atomic Energy Agency (2005) [5]

Of the ~1,200 radioactive isotopes currently known, only a few occur naturally. For example, uranium consists of two long-lived, weakly radioactive isotopes: U-238 (99.3% natural abundance) and U-235 (0.7%). Most importantly, U-235 is also fissile,[9] making it a useful source of energy. Standard nuclear fuel requires industrial enrichment of the U-235 content to between ~5% and ~20%, depending on reactor type. During U-235 fission, some of the excess neutrons are captured by the remaining U-238, transforming via U-239 and neptunium-239 to plutonium-239—the isotope used in modern implosion-type nuclear weapons.

Terrorist groups have repeatedly shown interest in acquiring radioactive or fissile nuclear materials, or active/retired nuclear weapons. Possession of either would give a small group an asymmetric capability when engaging with a state. Both fissile and non-fissile materials can be utilized in several ways: (1) combined with a standard explosive charge to form a radiological dispersal device (RDD); (2) hidden in a high-traffic public area as a radiological exposure device (RED); and (3) the direct contamination/poisoning of individuals, water and food supplies.

An [unlikely] success story is the security of nuclear weapons. Since August 1945, no state or group has detonated a nuclear or radiological device outside of planned, controlled testing. Eight nations have test detonated a total of 2,056 nuclear devices: 528 in the atmosphere and 1,528 underground. As of early 2021, the latest was an underground test by North Korea on September 3rd, 2017 [6]. There are an estimated 13,500 nuclear warheads in existence, made up of deployed, stockpiled and retired weapons [7].

The security concerns here are not just related to rogue-state proliferation and use. The portability of the common smaller nuclear devices, relative simplicity of some primitive nuclear weapons designs, ease of plutonium diversion from certain reactor types, or even outright theft of highly enriched uranium—all combined with a scientifically literate, well-funded and organised terrorist group—creates a realistic risk to the public from a stolen nuclear weapon or improvised nuclear device.

Nuclear Weapon Yields

In August 2020, 2,750-tonnes of ammonium nitrate fertilizer stored in a warehouse in the Port of Beirut exploded, leaving a crater ~250-m across, 204 people dead, 6,500 injured and ~300,000 homeless. The explosion registered as a magnitude 3.3 earthquake on nearby seismometers. The blast yield was equivalent to ~1,000-tonnes of TNT. For comparison, seventy-five years earlier, the *Fat Man* nuclear implosion device triggered over Nagasaki produced an equivalent yield of ~20,000-tonnes of TNT—from a 9-cm diameter, 6.2-kg sphere of plutonium. The estimated death toll was 74,000 people. The simpler *Little Boy*

[9] Capable of sustaining a nuclear chain reaction using thermal (slow) neutrons. U-238 is also fissionable but requires high energy (fast) neutrons to do so—conditions found only within an exploding nuclear warhead or fast-neutron reactor.

gun-type uranium device used three days earlier at Hiroshima, required 64-kg of uranium enriched to 80% U-235 to generate a yield of 15,000-tonnes of TNT. The blast killed an estimated 140,000 people, around 40% of the town's population.

The modern W88 thermonuclear (fusion) warhead used in the latest Trident II submarine-launched ballistic missile has a yield of 475 kilotons TNT. The W54, developed in the 1960's was a cylindrical 40-cm long, 27-cm diameter 'backpackable' implosion device, weighing only 23-kg and yielding around 10-tonnes TNT equivalent.

The UN *Treaty on the Prohibition of Nuclear Weapons* outlaws the creation, ownership and deployment of nuclear weapons by signatory states. It entered into force in January 2021—without any of the current nuclear powers signing up.

2 Risk

This section highlights the cradle-to-grave security risks[10] created by the worldwide prevalence of radiological and nuclear material linked to globally interconnected supply chains and diverse, complex national regulatory systems. We review historic system failures and explore the implications of societal, technological, and environmental change. Finally, we conclude by stressing the need for human-led, AI-supported decision systems to help users and regulators make sense of these rapidly evolving, multi-disciplinary security threats.

2.1 Threat Assessments

Radiological and nuclear security is underpinned by a complex mix of regulations and safeguards,[11] established under variable policy and economic constraints (Fig. 4). This basic process is used at all scales, from the IAEA's international role in ensuring safe, secure and peaceful cooperation in the use of radioactive materials, to the development of local industrial rules governing the use of radiological devices.

[10] We loosely define 'security risk' as any event(s) leading to unintended consequences. Events can be planned/unplanned, accidental/deliberate, benevolent/malicious.

[11] We consider the common role of 'safeguards' to include all radiological materials–not just fissionable nuclear materials.

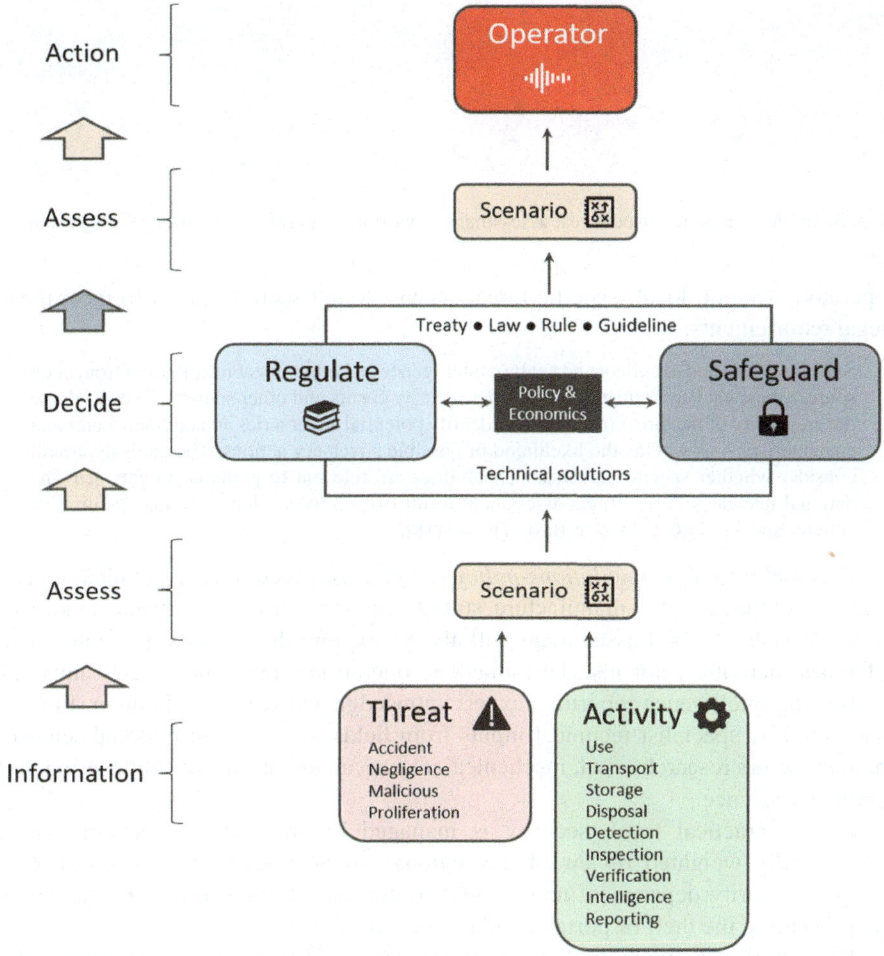

Fig. 4 Simplified process for development of radiological and nuclear materials controls. Can be applied at all scales, from international guidance (IAEA) to national regulations and law. 'Operator' includes any person or group with controlling responsibility for a radioactive material or device, e.g. the device owner, a user, facility where it is held, haulage driver, transport or storage company etc.

With an historic focus on counter-proliferation, the IAEA recommends member states use relevant, scalable threat assessments[12] to develop suitable national laws, regulations and safeguards.[13] Subsets of these assessments will then help facility

[12] A threat assessment determines the credibility and seriousness of a postulated threat—and its probability of occurrence.

[13] IAEA recommends regulators and operators follow the Design Basis Threat (DBT) approach to designing physical protection systems.

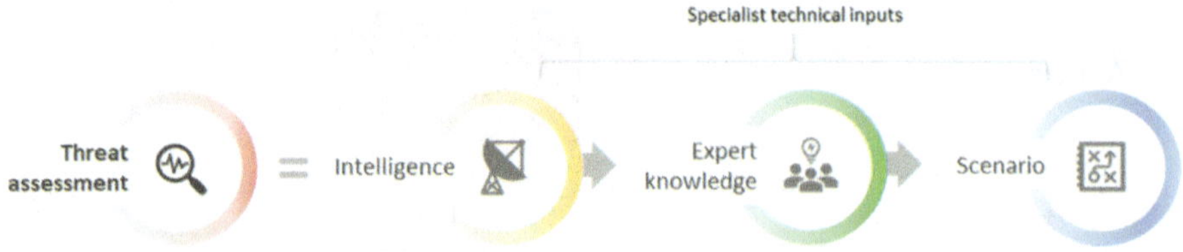

Fig. 5 Inputs to a generalised threat assessment supporting material and facility security plans

operators develop local, specific threat scenarios and security plans to meet these legal requirements.

> States are expected to collect and analyse intelligence and other threat information from open sources, past nuclear security events, other security events and other sources. They evaluate the credibility of the threat information, identify potential adversaries and their attributes and characteristics, as well as the likelihood of possible adversary actions. The analysis should consider whether specific adversary capabilities are relevant to potential targets. [..] The national nuclear security threat assessment result is used to develop material-, facility- or activity-specific DBTs [Design Basis Threats] [8].

Feeding into this *regulations-policy-safeguards* process is a detailed understanding of the risks[14] of manufacture, storage, transport, use and disposal of radioactive materials. A good assessment will always explore the broadest possible range of linked activities, not just the immediate operational use. The process involves combining intelligence reporting, expert knowledge and scenario planning (Fig. 5), supported by specialist technical inputs from fields as varied as: material science, human factors research, civil, mechanical, electrical and electronic engineering and computer science.

At the practical level, security is managed by the facility operator, based on a locally weighted review of the national threat assessment. For instance, a hospital security department may choose to concentrate their protective activities on preventing the theft of portable radiological devices.

Security threats typically derive from the actions of states, terrorists, extremists, criminals, contractors, suppliers and employees: whether intentionally, negligently or accidentally. In current security planning, the focus is usually on deliberate malicious acts leading to sabotage of a facility or theft of the radioactive material. Nuclear and radiological safety[15] issues are rarely considered at the same granularity. This artificial separation is unwelcome, as accidents[16] can create increased opportunities for malicious use, and vice-versa. The administrative loss of the Samut Prakan

[14] For our purposes, an event potentially leading to negative consequences.

[15] Safety can be defined as the establishment of proper operating conditions, prevention of accidents and mitigation of consequences, resulting in the protection of workers and the public from the dangers of radioactive materials. Safety is the responsibility of the facility operator, supervised by a national regulator.

[16] We define an accident, in the broadest terms, as an uncontrolled event caused by known, unaddressed, unrecognized or unforeseeable issues. For example, the Fukushima nuclear disaster can be viewed as *unforeseeable*: we cannot predict earthquakes and this was the largest ever recorded

teletherapy device (see text box) led to three deaths and many injuries. This could easily have morphed into serious malicious use were the orphaned source recognised and sold on to any of the regional Al Qaeda-affiliated groups active at the time.

A more robust approach is to combine both security and safety assessments under a single threat programme. Redefining the main issue ensures the process begins with the understanding that security threats can result from malicious activity, negligence *or* accidents:

> The primary radiological and nuclear materials threat to our health and environment stems from a **loss of control**. This can be the physical loss of a radioactive material via deliberate theft, negligence or accident; or a loss of functional control through failure in a human or cyber-physical system.

2.2 *Scenario Planning*

Operator-led facility scale threat assessments, whether security or safety focused, are designed and evaluated using scenario planning (Fig. 6).

A scenario is simply a description of possible actions or events that may occur in the future. By modelling the likely effects (results) and estimating their impact on the regulatory framework (health, environment, business, society), operators use scenarios to design and prioritise viable (and affordable) security measures.

Identifying event(s) for assessment is the key to successful security planning. An event captures an object's properties in space and time, and is mostly subjective. Often overlooked, the ***time*** function means an object is always being influenced by an event; it does not require any sudden or unusual action or change to act on the object to make it relevant. For example, the following is a measurable event:

Teletherapy device is locked in the radiology store between 09:00 and 11:00-hrs.

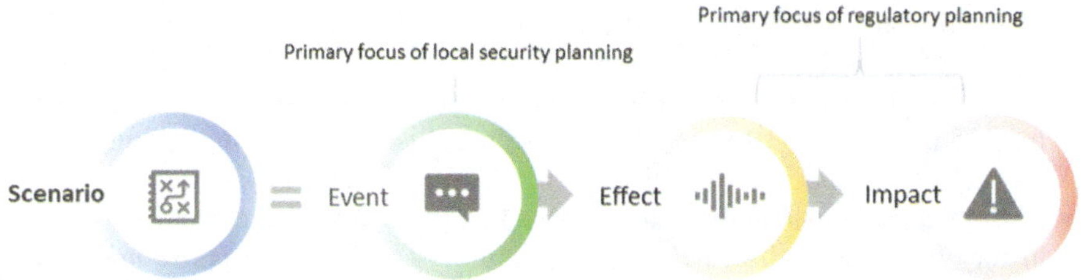

Fig. 6 Inputs to a generalised security scenario supporting material and facility security plans

in Japan; or as *unaddressed*, after action was not taken following an internal report indicating the plant's sea wall was built too low for the local tsunami threat.

This concept is important, as it creates a natural baseline for our primary task: identifying out-of-tolerance security-related events. While this principle applies equally to any threat assessment, radioactive decay adds a unique, changing material property that requires continuous consideration in any radiological scenario.

2.3 Human Understanding

While the 'capture every event' approach is important in identifying and managing complex threats, it does generate vast quantities of information. This can make human understanding difficult. As a result, scenario planning tends to ignore slow or steady state events assessed as lower risk, focussing instead on induced, rapid change events. These human bandwidth issues also lead to the frequent re-use of *ready-made* scenario questions, built on past experiences:

> How do we prevent another Fukushima Daiichi type disaster?

This early introduction of historic (experience) bias can lead to the hardwiring of common assumptions into the security plan:

- We assume the future will look much like the past.
- Scenarios are built around historic events.
- Where novel event types are considered, likelihoods are often weighted towards aspects of events that have occurred previously.
- Scenarios are often built within thematic and technical silos.
- Realistic multi-disciplinary scenarios create scalar problems for human understanding.

The Problem with History

Picture, in early 2001, using standard scenario planning to assess the broader terrorist threat. Law enforcement experience over the previous 10-years suggests a focus on protecting cities against car and truck bombs: 34 significant attacks on buildings, accounting for 1,458 deaths worldwide. Working in parallel, a transportation security team is focused on understanding aircraft hijacking: 200 incidents, most driven by local politics, accounting for 425 deaths over 10-years. Neither scenario is capable of modelling what subsequently happened on 11th September 2001 in New York City.

So, the problem remains—how do we link these knowledge silos, avoid the history trap, and realise the 'novel' question: "*Could terrorists use hijacked aircraft to attack buildings*?".

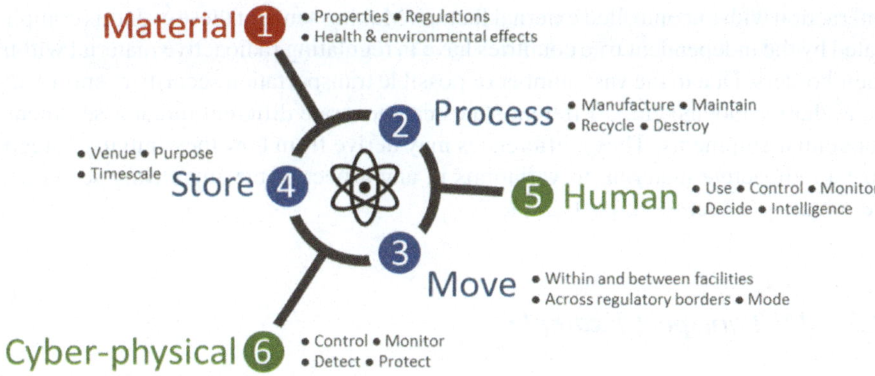

Fig. 7 Generalised life-cycle events impacting radiological and nuclear materials

2.4 Building Relevant Scenarios

Building and testing relevant, accurate scenarios, even at the local facility level, requires a review of the full lifecycle of a material. This means constantly collating and interpreting up-to-date information from six influencing factors (Fig. 7), grouped into three broad thematic areas: Material and device; Processing, Transport and storage; Human and cyber.

But what level of knowledge and scenario development is necessary—or achievable? Does a hospital security team need to know the origin of the cobalt in a teletherapy device? Perhaps—if it later becomes apparent the natural cobalt was obtained from artisanal miners using child labour. Reputational damage can quickly translate into physical and cyber protest. Even secondary, behavioural impacts should be included in a scenario impact assessment, for example: patients subsequently refusing to attend essential radiological treatment.

Any security incident will usually involve inputs from all three thematic areas, for example: *material* + *move* + *human* (transport incident), or *material* + *process* + *cyber* (manufacturing incident). Within these groupings, the most common driver of security incidents is related to human factors. For instance, in 2019, Australia found ~60% of their recorded RN incidents[17] were initiated by human error. Equipment-related issues accounted for a further ~20%, made up of software and hardware failures, alongside operator mistakes [9].

To illustrate the range and complexity of potential interactions between the six security factors, we consider the interplay between regulation and two sources of common failure: material transport and material responsibility/control.

In 2019, ~50% of worldwide reported radiological and nuclear incidents occurred while a material was being moved (data from the CNS Global Incidents and Trafficking Database [3]) Philosophically, any transportation event increases a material's

[17] Primarily related to medical imaging.

interaction with uncontrolled external factors. Making sense of these events is complicated by the independent role countries have in regulating radioactive material within their borders. Due to the vast number of possible transportation security events (accident, theft, sabotage etc.), different states can generate different threat assessments for similar shipments. These differences may derive from how they initially categorize a radioactive material, to variations in any aspect of the supporting scenarios, i.e. event → effect → impact.

2.5 RN Transport Example

The IAEA continues to push for a consistent approach, especially for the international movement of radioactive materials [10]. Even so, three different regulatory approaches are commonly followed:

(i) Prescriptive—national regulator specifies exact security measures required for material transport.
(ii) Performance—transport companies design and demonstrate their own systems to meet regulatory goals.
(iii) Combined—national regulator draws on elements of both prescriptive and performance approaches.

Spanning five countries, Fig. 8 illustrates a typical international supply chain supporting the development of a modern Co-60 teletherapy device. For simplicity, we focus on the cobalt and uranium source materials, while ignoring common ancillary items such as shielding components.

The life-cycle radiological (uranium concentrates and cobalt-60 metal) transport requirements for our example Co-60 teletherapy unit include multiple sea, rail, land and air journeys (Fig. 8 lower panel). However, it is not simply the physical protection of radioactive materials during transport that generates risk (mode, distance travelled, packaging etc.). A significant proportion of the threat derives from procedural failures and confusion over responsibility (see text box).

Nuclear Weapons Transport Incidents

In August 2007, a B-52 aircraft from Minot Air Force Base (North Dakota) was scheduled to transfer six unarmed (dummy warhead) AGM-129 ACM air-launched cruise missiles to Barksdale AFB (Louisiana) for decommissioning, a flight distance of 1,200-miles. Process and procedural errors at Minot led to the loading and transport of six missiles armed with W80-1 (5 to 150-kiloton) nuclear warheads. For 36-h, air and ground crews at both sites believed the missiles were unarmed. As a result, the aircraft and missiles were not subject to additional safety and security precautions required for nuclear weapons

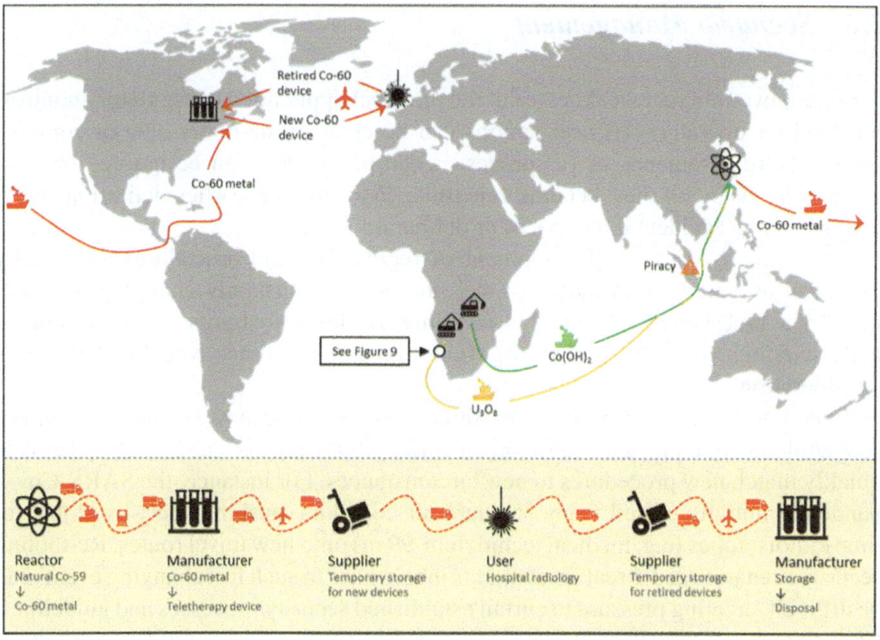

Fig. 8 Stylised international supply chain for a hospital Co-60 teletherapy device (primary natural components only). Understanding the life-cycle transport risks means considering all radiological aspects, including manufacture, supply, use and disposal. Green—not radioactive; orange—weakly radioactive; red—highly radioactive. The lower panel shows a more detailed view of the likely minimum transport events within the higher risk (red) cycle

handling, transport and storage. An investigation found multiple human failures to follow standard procedures, alongside serious gaps in supervision. The result was new policy directives, regulations, additional resourcing, improved training, mandatory checklists and surprise inspections.

The year before (August 2006), four classified nuclear weapon fuses used to trigger warheads above their targets, were mislabelled as helicopter batteries and shipped from the USA to Taiwan. The mistake was discovered 18-months later, and the parts returned. The investigation identified multiple contractual and procedural failures in a Defense Logistics Agency warehouse—leading to a cascading series of mislabelling, storage and shipping errors.

2.6 Scenario Management

There are two inherent weaknesses in the practical application of regulatory controls imposed on operators: (i) user actions can be error prone under new or complex regulatory requirements, or (ii) processes and procedures can be ignored or side-stepped by users as they become repetitive. As with our earlier definition, these actions can be accidental, negligent or deliberate.

Scenarios used to the development of regulations and procedures are usually modelled as one-off snapshots in time, not as live, constantly changing inputs to the threat assessment. Again, this is primarily down to limits of expert human understanding (mental bandwidth), funding and the time available for active consideration.

Recognising that scenarios and threat assessments may be moving out of scope/tolerance requires a continuous security review process, alongside systems to quickly match new procedures to new circumstances. For instance, the SARS-Cov-2 pandemic initially closed many standard air corridors, pushing short-lived just-in-time radioisotopes (e.g. medical technicium-99 m) onto new travel routes. Re-tooling security scenarios and threat assessments in response to such fast changing events can be difficult, creating pressure to curtail established security processes and guidelines: "*Just this once, let's allow..*".

Even where physical protective measures can be quickly re-assessed, interaction of a shipment with new, potentially inexperienced personnel along a temporary routing can still create significant uncertainty in security planning.

The management of the material or device is repeatedly fragmented as control[18] transfers between individuals and across tasks. Throughout a material's life cycle, responsibility can be repeatedly switched or shared between regulator, owner, user, transport and storage companies, health and safety teams, local security, law enforcement and others. Figure 9 illustrates how responsibility for a uranium concentrate shipment at a seaport can vary throughout the earliest stages of the uranium cycle.

Misunderstandings over responsibility have played a key role in many of the incidents discussed in this chapter. Improving our understanding of responsibility is essential in building relevant, robust scenarios and threat assessments; the smallest changes to the scenario can significantly alter the subsequent threat assessment. For example, in the figure above, forklift operator #304 has direct control of the uranium concentrate during transport across the dockside (Event R-2–7). This is followed by a handover of direct control to crane operator #322 (Event R-2–8). Imagine then, during the subsequent lift from dockside to cargo hold, the load parts from the crane and falls from height onto the ship.[19]

---

[18] We define controlling responsibility as taking four forms (i) 'Command' or organisational oversight; (ii) 'Supervisory' or general oversight; (iii) 'Custodial' or specific oversight; (iii) 'Direct' or immediate physical control.

[19] This scenario draws on events at the Port of Halifax in March 2014. Incorrect attachment of a shipboard container to a shoreside crane led to 4 cylinders of fissile low enriched uranium hexafluoride, weighing 18-tonnes, being dropped into the ships hold from a height of 7-m. Although the

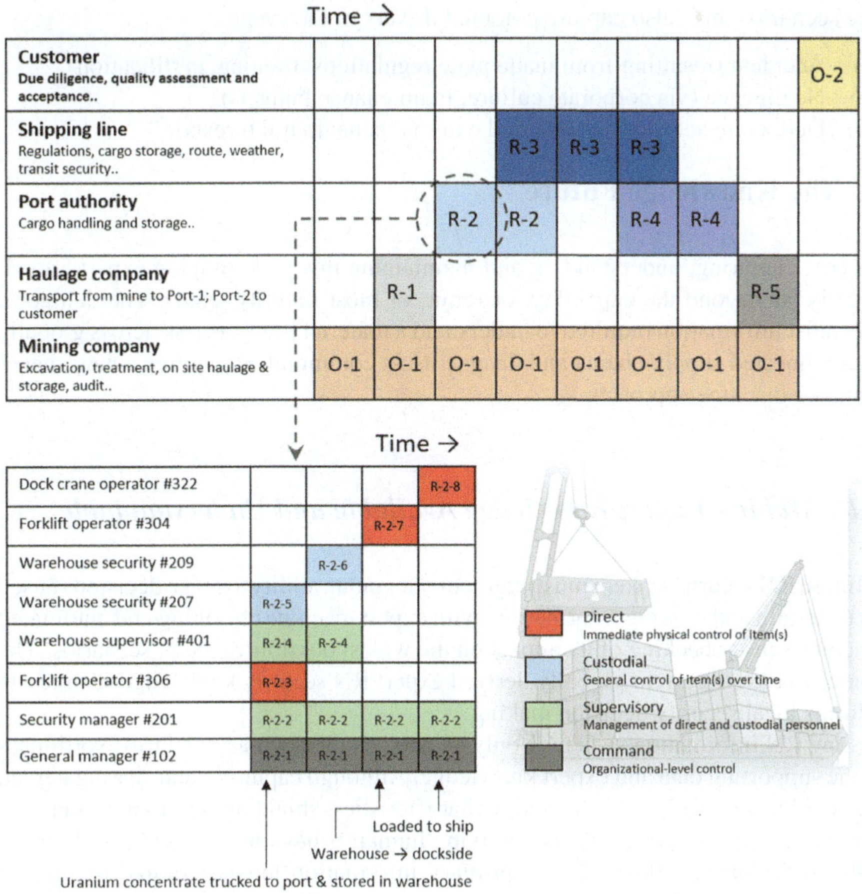

Fig. 9 Idealised responsibility/control chart for movement of uranium concentrate at originating seaport in Fig. 8. Example assumes the customer has no responsibility until they accept delivery. The idealised information granularity will include an understanding of every responsible party's training, skills and experience. O = ownership responsibility; R = controlling responsibility

For a security team to anticipating this event, they must understand, design and review multiple scenarios covering the entire operation:

(i) Training and supervision of anyone with responsibility for—or control over—the material.
(ii) Human factors (tiredness—loading at the end of a shift versus beginning).
(iii) Maintenance of the lifting equipment, containers and attachments.
(iv) External factors impacting the material, such as operational weather limits.

---

cylinder protection features withstood the fall, the container port was closed for 24-h while clean-up and site investigation took place.

The scenarios must also capture potential drivers of the event:

(i) Accident (resulting from inadequate regulations, training, certification).
(ii) Negligence (via corporate culture, maintenance budgets).
(iii) Deliberate acts (e.g. insider and external behavioural threats).[20]

3 The Knowledge Future

Clearly, capturing, understanding and maintaining this wide-ranging expert knowledge is far beyond the capability, or remit, of most security teams. The deluge of specialist information required to understand a material lifecycle risk across globally interconnected supply chains and diverse users communities requires a new shared human—machine approach.

3.1 Making Expert Knowledge Available and Understandable

A future RN security system must augment our human ability to make decisions based on uncertain and missing information, with expert machine reasoning and automated machine sense-checking of its impact on the widest possible range of scenarios. The goal is to ensure current globally derived expert RN security knowledge seamlessly informs local security decision making.

Any RN information system is only as good as the accuracy and trustworthiness of the supporting data and expert knowledge. Although capturing and sharing agreed measurable knowledge (e.g. material characteristics, shielding requirements etc.) is straightforward, mapping inputs involving human behaviour are significantly more difficult. Figure 10 illustrates the primary information inputs required to support development of such a system.

In this example, building an RN knowledge graph[21] creates a single, globally shared RN vocabulary, while also capturing the key relationships underpinning RN security best practice. A local security team can then map their own facility data to this 'live' expert knowledge graph, allowing them to compare their approach with current best practice. Capturing these actions digitally also creates opportunities for semi-automated machine reasoning to continuously sense check the local facility plan against the wider expert knowledge base. Figure 11 illustrates this approach with a generic knowledge base (A), subsequently populated with graph data by a local security team (B).

---

[20] Although this scenario is focused on human—mechanical events, the same process is followed when considering cyber—physical events.

[21] In computer science, knowledge graphs are a mechanism for adding context to otherwise 'raw' information and data. The resulting graph essentially maps human-held expert understanding about the relationships between 'things' and makes them machine-readable. This provides a framework for machine reasoning, data integration, unification, analytics and sharing.

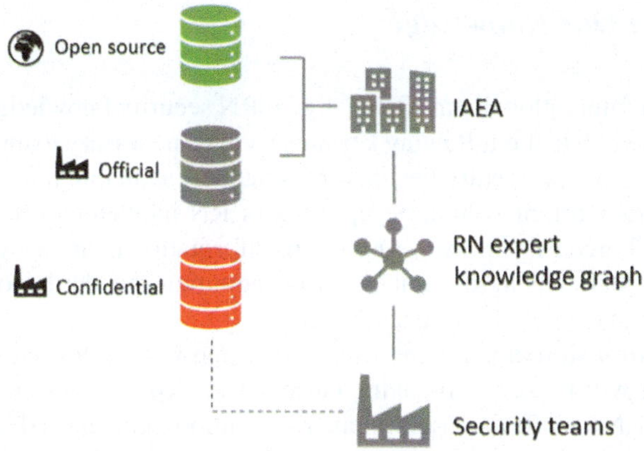

Fig. 10 A future human—machine knowledge approach to managing radiological and nuclear security. Countries and individual facilities share relevant releasable information on materials and processes with, for example the IAEA, who capture and maintain this global learning within an expert radiological & nuclear information management system (see Fig. 11). Where countries or facilities hold additional sensitive knowledge unsuitable for wider sharing, this can be managed as an internal, secure overlay within their locally held copy of the general open access knowledge graph

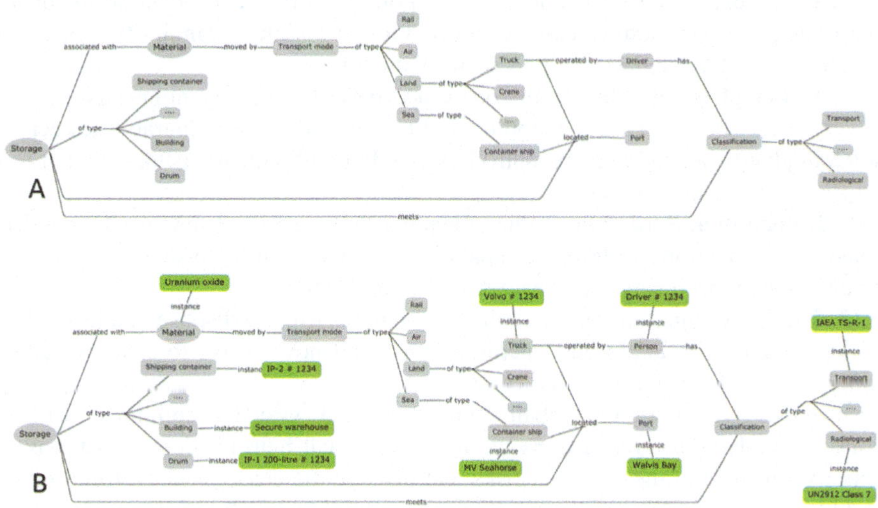

Fig. 11 Illustrative example of a knowledge graph based around the storage and transport of RN materials. **A** shows the generic relationships (the ontology); **B** shows the addition of local data (the full knowledge graph)

3.2 Gaps in Our Knowledge

In supporting a future globally available digital RN security knowledge base, further activity and research in the following key areas will be necessary to ensure any future information system continually benefits from the best available information:

1. **Reporting**. Current voluntary reporting of RN incidents to the IAEA generates opaque and inconsistent data. Any future information management system will require an international agreement covering common standards for reporting of events involving category 1 and 2 materials.
2. **Information sharing**. Secure, trustworthy and sometimes anonymised information sharing will be key to breaking current knowledge siloes and subsequently populating the RN knowledge graph. Here, recent improvements in distributed ledger technologies and homomorphic encryption will ensure the widest possible roll out of accurate, real time querying of sensitive data.
3. **Behaviour**. Perhaps the most difficult area to understand and map will be the role of human failures in RN incidents. Deliberate or accidental human actions are cited in over 60% of incidents reported to the IAEA. This will become more important as employers continue the move towards higher staff turnover, contractor-based arrangements in many RN industries.
4. **Transport**. Since around 50% of all reported RN incidents occur during material transit, it is essential to capture information at a higher granularity to support development of a detailed transport security model.
5. **Cyber-physical**. The likelihood of a cyberattack on any connected system is very high (100%). Timely mapping and understanding of vulnerabilities across software, hardware and human controllers—and the subsequent visualisation of this information, remains essential.
6. **Detection and forensics**. When planning fails, security teams need an understanding of detection and forensic capabilities. As with cyber-physical, both are fast changing areas requiring specialist technical knowledge.
7. **Policy**. As international manufacture—supply—use—dispose chains evolve, policy makers, regulators and operators must find novel ways to share accurate, timely views of developing security issues.

Increasingly complex, globally interconnected systems combined with a constantly adapting threat means we need additional help to identify patterns of activity, understand the resulting risks and manage the consequences. Building on technical improvements in the seven areas listed above, hybrid human—machine systems capturing explicit expert knowledge, mapped to the latest human behavioural modelling will become the foundation of this process.

Development of this global, shared radiological and nuclear information management system is essential to giving policy makers, regulators, operators and security teams access to the timely, accurate, ever-changing knowledge they need to protect us.

References

1. https://www.researchandmarkets.com/reports/5140237/nuclear-medicine-radioisotope-global-accessed. Accessed April 2021
2. https://world-nuclear.org/information-library/non-power-nuclear-applications/radioisotopes-research/radioisotopes-in-medicine.aspx. Accessed April 2021
3. https://www.nti.org/analysis/articles/cns-global-incidents-and-trafficking-database. Accessed Feb 2021
4. https://www.www.iaea.org/nuclear-security/incident-and-trafficking-database. Accessed March 2021
5. International Atomic Energy Agency (2005) Categorization of radioactive sources. IAEA Safety Standards Series No. RS-G-1.9. IAEA, Vienna, p 70
6. https://www.armscontrol.org/factsheets/nucleartesttally. Accessed July 2020
7. Stockholm International Peace Research Institute (2020) World nuclear forces. In: SIPRI Yearbook 2020: armaments, disarmament and international security. Oxford University Press, p 720
8. International Atomic Energy Agency (2009) Development, use and maintenance of the design basis threat. IAEA Nuclear Security Series No. 10. IAEA, Vienna, p 30
9. Australian Radiation Incident Register (2021) Australian radiation incident register report for incidents in 2019. Australian Radiation Protection and Nuclear Safety Agency, Australian Government, p 31
10. International Atomic Energy Agency (2018) Regulations for the safe transport of radioactive material, IAEA Safety Standards Series No. SSR-6 (Rev.1). IAEA, Vienna, p 190

References

Explosives and Explosive Effects

W. G. Proud

1 Global Overview

Explosions are generally reported in the media with a mix of tragic content and apocalyptic detail. In many mass casualty incidents, explosive devices are often present in a primary or secondary role. The term 'gas explosion' generally provokes thoughts of a gas cooker or gas cylinder exploding with an image of a damaged building and reports of a limited number of deaths and injuries. Explosions from natural phenomena are the rarest, however cyclones and storms can produce explosion-like effects. It is the aim of this chapter to cover the key points and function of an explosive, the immediate effects on people and on urban structures. There is an extensive and well-developed literature in each of these areas which can be accessed via the suggested reading and the references.

1.1 Deliberate Acts

Addressing the threat of terrorism has been a long-standing priority of the international agenda, particularly following the events of 9/11. The twenty-first century has seen an upsurge in terrorist activity. Terrorism remains the subject of intense analysis by a broad range of research disciplines: studying changes in terrorist tactics, recruitment, operational scope, ideology and weaponry. Though attacks may vary greatly in their execution, from lone wolf to coordinated incidents, a hallmark of modern terrorism has been psychological warfare achieved through indiscriminate and ostensibly random attacks on the public with the intention of inflicting maximum morbidity and mortality. For this purpose, Improvised Explosive Devices (IEDs) have

W. G. Proud (✉)
Blackett Laboratory, Institute of Shock Physics, Imperial College, London SW7 2AZ, UK
e-mail: w.proud@imperial.ac.uk

P.D.E. Biggins and D. Chana (eds.), *CBRNE: Challenges in the 21st Century*, Advanced Sciences and Technologies for Security Applications,
https://doi.org/10.1007/978-3-031-17374-5_5

been utilised globally with devastating effect. A range of epidemiological studies exist reporting blast injuries sustained by military populations [1] with relatively less focus on the civilian environment. This chapter will focus on the civilian environment.

Bombings remain the leading worldwide cause of civilian fatalities due to terrorism. This demands an understanding of modern terrorist bombing trends to inform mitigation strategy. The objective of the study by Magnus et al. [2] was to identify the occurrence and severity of bombings against civilian targets in diverse attack settings, and to establish corresponding blast injury profiles. Data was obtained from analysis of the Global Terrorism Database (GTD) and a meta-analysis of blast injury data derived from the PubMed database. These databases are constantly updated and are widely accessible.

In order to link the casualties to injury mechanism and scenario, Tables 1 and 2 give a summary of types of injury seen and the scenario types. Further details on the types and causes of injury form the second section of this chapter. In the medical literature the types of blast injury are classified into four groups: primary are those injuries associated with the blast wave alone; secondary are those due to fragments or projectiles thrown by the blast; tertiary are due to the human body being thrown against other objects; quaternary include burns, toxic shock and acts as a catch-all class for injury types which are not mechanical in nature.

Figures 1 and 2 show the growth in reported incidents in the period 1970–2018. Closed environment explosions were associated with significantly greater ($p < 0.05$) mortality than open spaces. Injury profiles were found to be significantly influenced by attack setting, with higher rates of primary injury on trains and buses, and secondary injury in open space.

For blast mitigation research, the civilian setting is of special interest due to victims largely being unequipped with Personal Protective Equipment (PPE) and

Table 1 Blast injury classifications for review of epidemiological data

	Blast injury classifications			
Blast injury mechanisms	Primary	Secondary	Tertiary	Quaternary
	Blast lung injury (BLI)	Penetrating torso	Skull fracture	Burns
	Tympanic membrane perforation (TMP	Penetrating head, neck and face	Other fracture	
	Intra-abdominal rupture	Penetrating extremity	Traumatic amputation	

Table 2 Scenario attack types

Mobile	Static locations
Train, trolley bus, subway trains Bus, van, vehicle, taxi, airplanes	Semi-confined space (SCS) such as Restaurants, banks, hotels, shopping malls, subways, etc. Open space (OS) such as Urban areas, villages, roads

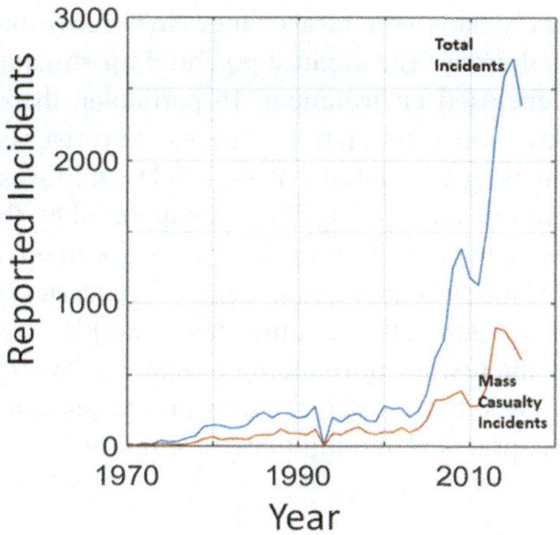

Fig. 1 Terrorist activity during the period 1970–2016. *Credit* W.G. Proud

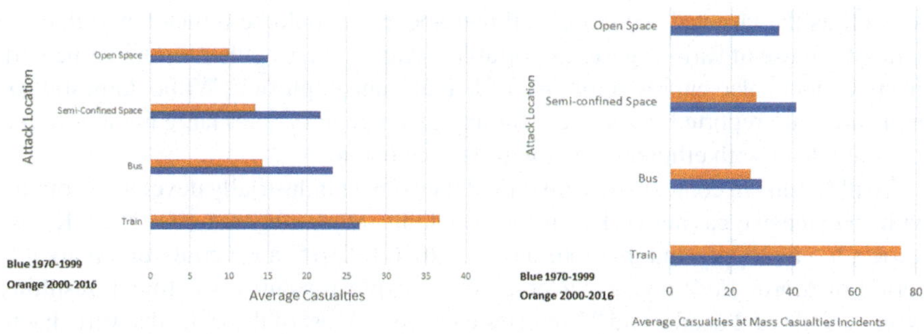

Fig. 2 Distribution of terrorist events across attack settings for all incidents and mass casualty incidents. *Credit* W.G. Proud

the need to find solutions which do not impinge on normal life. Patients will often experience a varied medical response and have varied transit times and manners to trauma centres. In terms of clinical blast injury data, different trauma registries or investigators provide wide-ranging degrees of medical record detail. The effect of the blast environment was examined by determining the severity of these events according to attack setting [2].

The large casualty population included in the meta-analysis enables conclusions to be drawn that may assist with informing blast mitigation strategy. Enclosed spaces are associated with enhanced primary blast injury, placing greater emphasis on the need to mitigate the blast wave. In open space, protection from fragment impact must be prioritised. Mitigation material design is fundamentally different in both cases. Blast wave mitigants may incorporate a deformable component for impulse energy absorption.

In the case of secondary injury, armour is conventionally designed for defeating ballistic or penetrating threats, typically utilising combinations of hard ceramic plates and aramid fibres. For civilians, structural changes to the environment must be sought rather than personal solutions. The greatest potential for structural design to provide mitigation is in the enclosed environment. In particular, there is a need here for effective primary blast injury mitigation. Though terrorist activity has increased across all attack settings, a key finding of the GTD study was the upsurge of bus and train attack frequency and severity. This is supported by the blast injury meta-analysis, with train incidents resulting in the greatest morbidity. In response to this, O'Neill et al. identified the structural components of trains and rail systems that must be addressed for mitigating the effects of the blast wave [3]. Areas such as the walls, doors and roof are of interest for introducing mitigation layers. An analogous need also exists in buildings where retrofitted cladding may be preferred. Such applications demand cost-effective, practical and lightweight solutions.

1.2 Domestic Explosions

As well as the targeted scenarios outlined above it should be remembered that the widespread use of various gases as portable forms of energy either through pipework or in cylinders also entails some risk of leakage and explosion. While domestic gas explosions are reported in the press for any given country, a far large number of gas leaks are dealt with efficiently and without explosion.

To place this in context some historic data from an industrially developed country with an extensive gas network is useful. Looking at the safety record of the UK with respect to gas supply and distribution from 2010 to 2015, a generally unremarkable period of 'normality': a yearly average of 25 explosions/fires was found, resulting in an average of 2 deaths and 32 injuries each year. Most of these deaths were due to the resulting fires and building collapse rather than the initial explosion. There was a significantly higher incidence of carbon monoxide poisoning resulting in an average of 7 deaths a year and an average of 290 incidents of non-fatal carbon monoxide poisoning [4]. Overall, the fatality rate from gases such as methane is very low given there may be up to 500,000 gas leakages reported in any year. Technology to prevent gas being released from heating and cooking appliances without being ignited, plus cut-out systems to switch off the gas if CO is produced, have had a major effect on these numbers. Similar data from the 1970s show hundreds of deaths per year due to carbon monoxide alone.

1.3 Industrial Activity and Explosive Storage

Industrial explosions can be more devastating in terms of amount of materials involved, infrastructure destroyed, and lives lost. In some cases, this may be due to fires during transport of flammable, volatile, or explosive materials. The following examples show the variety and scale of the effects seen.

The Bhopal disaster in India occurred due to pressure developing in a storage vessel, leading to an explosive gas release followed by further gas venting in December 1984. This release of methyl isocyanate led to the immediate death of between 3787 and 16,000 people while ~500,000 are said to have suffered ill effects.

The Ryongchon Station disaster of April 22nd, 2004, occurred when a flammable cargo exploded in North Korea, near the border with the People's Republic of China. According to the Red Cross, 160 people were killed and 1,300 were injured in the disaster. A wide area was reported to have been affected, with some airborne debris reportedly falling across the border in China. The Red Cross also reported that 1,850 houses and buildings had been destroyed and another 6,350 had been damaged.

Lethal fires in explosive storage facilities or factories occur, worldwide, on average once every two years, with many other fires not causing immediate harm. This relative safety is due to the strength, remoteness and security of explosive factories and storage facilities, the use of safety-distance rules, lightning protection, and specialist handling capabilities. However, effective stewardship and control of explosive materials is needed: an example occurred on 11 July 2011, when a large amount of ammunition and military explosives that had been stored outdoors for over two years at the Evangelos Florakis Naval Base, Cyprus, self-detonated. This killed 13 people, including the Commander of the Navy, the base commander and six firefighters. A further 62 people were injured. The ammunition had been seized in 2009 from a cargo ship bound for Syria. The explosion severely damaged hundreds of buildings including the nearby Vasilikos power station causing widespread disruption in the supply of power to the island.

In 2020, a massive explosion devastated the port area and much of the surrounding zone of Beirut [5]. This was attributed to the combustion and then detonation of over several thousand tons of impounded ammonium nitrate intended for use as a fertiliser. This material had been impounded over six years previously but had no effective stewardship by the Lebanese authorities.

Quarrying and mining account for approximately 99% by mass of all explosive use in the world. The petrochemical industry uses explosive charges in exploration, in perforating oil and water bearing rock, and in general blasting. This is not without hazard, and incidents related to explosion occur, though in the context of mining the main dangers are the collapse and flooding of tunnels, accidents with vehicles and machinery, plus long-term effects of dust and exposure to toxic materials.

1.4 The Natural World

The majority of the explosive incidents reported in the natural world come from two scenarios involving the rapid release of energy and gas expansion. Firstly, there are volcanic eruptions, where sudden release of pressure can produce blast waves and pyroclastic flows. Pyroclastic flows are relatively fast-moving mixtures of hot gas and fragments. Volcanic activity is a well-documented phenomena and humans have tended to settle near active or dormant volcanoes due to the fertility of the soil. The second natural phenomena that is dramatically explosive is the impact of asteroids, meteorites, and comets. With impact velocities of up to 70 km s^{-1}, the contact between the incoming body and the planet results in intense localized heating, vaporization, and explosion with fragmentation. Such impacts are associated with the death of the dinosaurs, a subject beyond the present chapter, and the explosion in Tungushka, Siberia in the early 20th Century. At this point there is no historical record of such impacts on urban centres or immediate societal change. However, longer term effects due to dust being thrown into the atmosphere by volcanoes and as lightweight impact debris has led to cooling, reduced plant growth, and starvation affecting a wide diversity of life forms.

If the phenomena of the natural world are classified by effects and not cause, then explosive-like effects are produced by tornados, whirlwinds, and hurricane conditions. The extreme winds produce strong loading on structures and objects are entrained into the flow. Secondary injuries are produced by being thrown or violently carried by the wind, while tertiary injuries from projectile impact are relatively common. Unlike explosive loading, which is generally millisecond to second-long, the loading from extreme weather can last minutes to hours leading to building responses ranging from sudden demolition to slow progressive collapse. For buildings in tornado areas any added protection in terms of shielding, construction regulation, and removal of objects and external features which can be moved by the wind along with the presence of internal safe spaces assist in producing more blast resistant structures and lower casualty rates.

2 Explosives and Blast: Kinetic Effect

In broad terms an explosion is a term used to describe the rapid expansion of gas. It may be from the rupture of a pressure vessel such as a gas cylinder, the sudden vaporisation of a liquid such as with water exposed to hot metal, or by a rapid chemical reaction as seen in an explosive. An extensive literature exists in this field and readers wanting to understand more fully should consult the texts by Cooper, Fordham and others [6–9]

Explosives form part of a range of materials classified as "energetic materials". Other members of this class include propellants and pyrotechnics. The distinguishing feature of energetic materials compared to other grouping of materials is in the much higher rate of energy release.

The basic chemical components involved in an explosive are a fuel, an oxidiser, and a material that allows a rapid ignition of the material. In terms of total energy released, energetic materials are not particularly distinguished from other chemical reactions; petrol and butter release more energy per molecule when oxidised than trinitrotoluene (TNT). This difference is that while petrol needs to be mixed with air, then set alight, the explosive comes with the fuel and oxidiser intimately mixed, often with both fuel and oxidiser present within the same molecule.

High Explosives, correctly called High Order Explosives, involve materials like TNT, HMX, RDX, and PETN often mixed with other chemicals which make the explosive composition more stable, easier to industrially process, or able to fit into cavities within a munition. When these materials detonate, they release their energy due to a shock wave being produced within them. A shock wave is a high-pressure pulse which moves through the material at supersonic speed. In the case of a detonation, the shock wave consists of a thin, often sub-millimetre, region where the explosive turns from a solid or liquid into a hot high-pressure gas (Fig. 3). The velocity of a detonation wave is of the order 8 km s^{-1} and the energy release rate is of the order of Gigawatts, i.e. the same output as a large electrical plant. For most explosive charges, depending on size, this energy release takes place over the time of microseconds to milliseconds. The pressure associated with detonation waves are of many hundreds of thousands of atmospheres. The produced gases expand quickly. A rule of thumb gives the velocity of product gas expansion as ~1/4 of the detonation velocity. It is not surprising that such aggressive energy release can be used to shatter and push materials at high velocities.

Within the class of high explosives there is a sub-division between so-called "ideal" and "non-ideal" explosives. Ideal explosives have very thin detonation-reaction zones where the reaction takes place on a sub-microsecond basis. The materials tend to have a shattering effect, brisance, on materials placed in contact

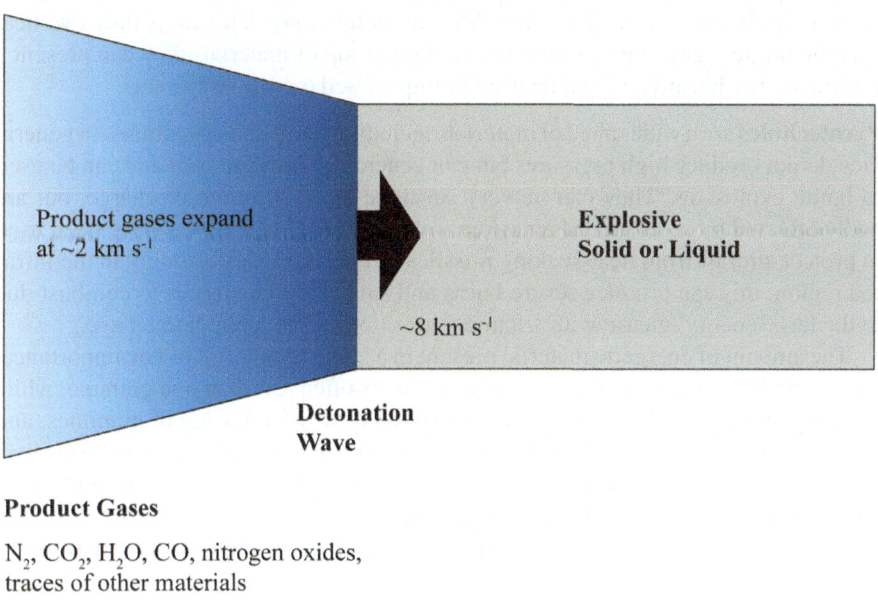

Fig. 3 Schematic of detonation process. *Credit* W.G. proud

with them. This class of materials includes TNT. Non-ideal explosives have much thicker reaction zones. The most widely used explosive in the world, "Ammonium Nitrate -Fuel Oil" (ANFO) falls into this class. The chemical composition of ANFO results in much lower pressures of detonation, a tenth and a half, of that from high explosives, producing less shattering effect. However, ANFO does generate a lot of product gas, giving the mixture a lot of 'heave'. The low detonation pressure means this mix produces fewer cracks, a lower brisance, in rocks however these cracks are then pushed open, heaved, by the detonation products. These properties make these materials particularly useful in the mining and quarrying industry.

Low Order Explosives are materials that give out their energy as a result of rapid burning, called deflagration. This class includes gun-powders where the reaction rate is much lower, of the order of a thousandth of that seen in high explosives. The total energy output is similar or higher than high explosives.

Both high and low explosives may need a degree of mechanical confinement in order to explode or increase the explosive effect. A gramme of gunpowder on a bench will react quickly with a flash of light, a small fireball, some heat but little other effect. The same quantity placed inside a sealed metal can where the confinement allows the hot gases to stay close to the powder, allowing pressure and temperature to build-up, accelerates the reaction until the confinement shatters and an explosion is produced.

Propellants are a wide group of materials where the reaction pressures are of the order of thousands of atmospheres and the reaction timescales are measured in milliseconds. These materials are used in rocket motors, to drive bullets and shells. In some cases propellants, if confined or impacted at high velocity, may detonate. This can be an issue in transport and storage. Missile motors may detonate if they overheat or if the product gas cannot vent quickly. As a group of materials they can present a significant fire hazard and can be used in improvised explosive systems.

Pyrotechnics are a wide range of materials including flares and obscurants. In general they do not produce high pressures but can generate significant heat and can be used to ignite explosives. They can be very sensitive to electrostatic discharge, but are badly affected by moisture. Magnesium-Teflon-Viton (MTV) flares have been used to protect aircraft from heat-seeking missiles as they emit very strongly in the infra-red region; this can produce severe burns and cause other materials to combust due to the large energy release with a flame that is not visible to the human eye.

The amount of energetic material present in a given scenario is of key importance. The propellant contained in a small arms round is often less than one gramme, while a hand grenade would contain something of the order of a few ten of grammes, and a large anti-tank mine would contain up to twenty-five kilogrammes of explosive. Fireworks storage units may contain tonnes of pyrotechnics, sometimes within the confines of a city that has built up around them.

Given this confusing array of systems, it is easiest to think of the munition or explosive scenario in terms of the explosive train (Fig. 4).

A variety of stimuli can be used to put the explosive train in motion such as standing on a mine, closing an electrical switch, activating a magnetic action, or a

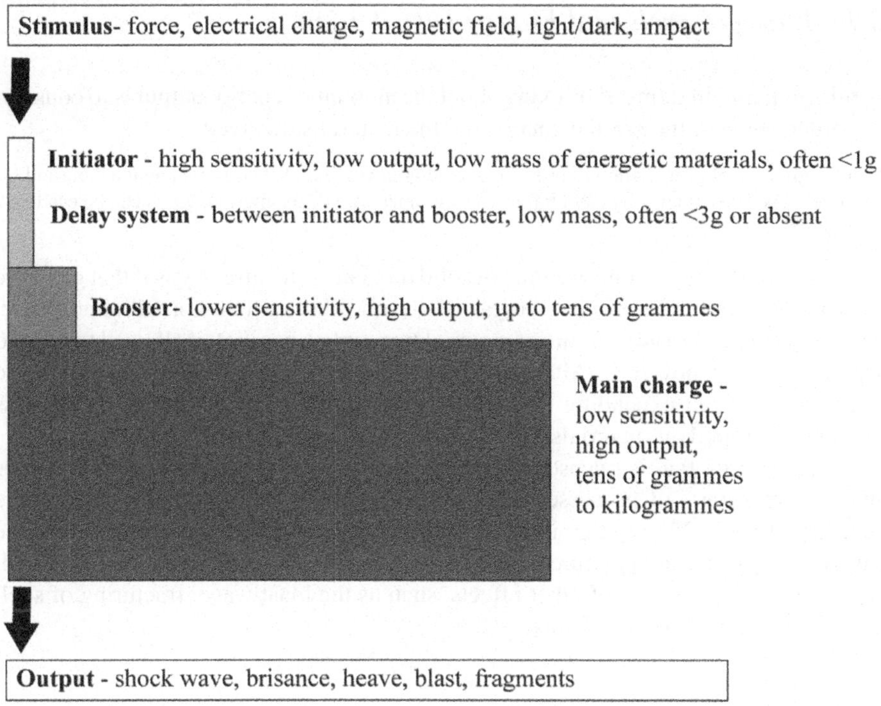

Fig. 4 The explosive train. *Credit* W.G. Proud

fire, etc. These actions input energy into an initiator. The initiator contains a material that is highly sensitive, combusting with ease. The main function of the initiatory system is to produce heat or a shock wave. In anti-personnel mines, a common design is based on a simple crush switch where the pressure stabs a metal pin into a small metal thimble filled with a sensitive explosive.

The booster system consists of an explosive compound that is less sensitive than the initiatory compound, but which will detonate due to the heat of shock waves from the initiator. The energy output of the booster is much higher than that of the detonator. The shock wave then transmits to the main charge.

The main explosive charge consists of a relatively low-sensitivity explosive composition that will detonate from the energy delivered by the booster. The mass of such main charges is extremely variable, however, the output from the main charge can be defined in terms of the outputs.

These outputs take the form of (a) shock waves transmitted into the immediate environment around the explosive, (b) the shattering of surrounding material, such as a metal case or rocks and soil, (c) the expansion of hot product gases, (d) the coupling of some of the explosive energy into the air, producing a blast wave, and (e) the aggressive acceleration of fragments, shrapnel or target vehicles and people.

2.1 Energy Levels and Energy Distribution

A rule of thumb in terms of thinking about the amount of energy output is to consider the order of magnitude of the energy and the materials involved.

> One gramme of propellant will produce one kilojoule of energy. Gases are about a thousand times less dense than solids and liquids. The temperature of the product gases can be easily 3000 K.

The result of this is that one gramme of solid turns into one litre of gas if that gas were at room temperature, however, the gas is ten times hotter than normal atmosphere. This means that the one gramme of material turns into a gas that, if allowed to expand freely, would occupy 10 L. Alternatively it would take ten thousand atmospheres to keep the gas from expanding. This order or magnitude calculation is for the less aggressive propellant materials.

Of the energy that is released by the energetic material the proportion that goes into different parts of processes is revealing. Again, a simple order of magnitude indicates that 20–30% will go into the kinetic energy of the fragments, 60% into the kinetic energy and temperature of the product gases, while the remaining 10% will be spread over a number of other effects, such as the blast wave, fracturing of shell casings, the motion of the ground, etc.

2.2 Formation and Velocity of Fragments

Many munition systems have a casing, usually of metal or plastic, often with shapers to form specific fragments. Some systems have very thin casings: this is generally done to reduce the metal content of the munition, make it harder to detect. This also has the effect of increasing the blast effect as less energy is used in fragmentation and acceleration of fragments.

The initial effect of the detonation wave is to send a shock wave into the casing and produce a violent acceleration of the casing [10]. The detonation products send a shock wave into the cylinder from the inner surface to the outer surface. A wave reflection takes place at the outer surface, sending a release wave back into the casing, towards the inner surface. This in turn is reflected from the inner surface as a shock. As this back-and-forth wave reflection process takes place, each reflection represents the acceleration of the casing expansion in a series of steps. As the casing is accelerated outwards it expands and becomes thinner, the outer surface moves from rest to reach a velocity of 0.8 mm μs^{-1} (corresponding to 800 m s^{-1}) within 5 μs. A more detailed discussion of how this can be calculated can be found in the specialised texts of the explosives engineering community, detonation symposium, and shock physics literature. The total distance moved by the casing within this very narrow time window was only 0.6 mm. Thus, the metal casing is subject to very violent acceleration. This causes fracture and fragmentation within the casing, producing metal splinters moving at high velocity.

The study of fragments and fragmentation took a major step forward during the Second World War (1939–1945), with the scientific efforts placed in the service of militaries of industrialised nations. The two major steps are named after their main originators: Neville Mott and Ronald Gurney.

The number and size of metal fragments from a casing was addressed by the Mott fragmentation criteria. This is a statistical model based on the idea of a rapidly expanding ring of metal which contains imperfections that form the basis of fractures which eventually break the ring, As the ring breaks the stress inside the resulting fragment reduces and further fragmentation does not occur. This model was developed and populated in a semi-empirical fashion, the results of theory and experiment were compared [11].

This model has been subject to development since the 1940s, with much recent effort being given by Kipp and Grady in the USA [11]. However, the original theory and Mott criteria are still regarded as valid and useful tools. Mathematically the fragmentation can be represented as:

$$N(m) = \frac{M_0}{2M_k^2} e^{-\left(\frac{m^{1/2}}{M_k}\right)} \tag{1}$$

$N(m)$ is the number of fragments that are larger than mass (m), m the mass of a fragment, M_o the mass of the metal cylinder, M_k is called the distribution factor which can be calculated from the following equation:

$$M_k = Bt^{\frac{5}{16}} d^{\frac{1}{3}} \left(1 + \frac{t}{d}\right) \tag{2}$$

where B is a constant for the particular explosive-metal pair that is used, t is the cylinder thickness and d is the inside diameter of the cylinder. From this a series of curves giving fragment size distribution can be generated [6].

The research of Gurney [12] considered the balance of energy in the system, how much of the released energy would be captured by the metal casing, and so predicts the expected fragment velocity. In the derivation of the velocity, the casing was assumed to remain intact. The shape of the explosive charge and casing has a major effect. In the equation below, the expression is for the velocities achieved by explosive loading of a cylinder:

$$\frac{V}{\sqrt{2E}} = \left(\frac{M}{C} + \frac{1}{2}\right)^{-1/2} \tag{3}$$

C—the mass of the explosive charge

M—The mass of the accelerated casing

V—Velocity of accelerated flyer after explosive detonation

$(2E)^{1/2}$- is the Gurney Constant for the explosive, it has the same units as velocity and is sometimes called the Gurney velocity. Each explosive has a particular value of Gurney constant and so can be thought of as ranking the ballistic effectiveness of a given explosive.

These two approaches may seem contradictory, one assuming fragmentation and the other an accelerated but intact material, but they are complimentary due to the timescales of the initial acceleration and fragmentation. The energy delivery by the detonating material occurs very quickly giving the initial impulse, while fractures take time to develop and open sufficiently to allow the product gases to escape. The combination of these two approaches provide a reasonable and predictive approach to fragmentation and fragment velocity.

2.3 Blast Waves

As indicated earlier, the product gases have a velocity of approximately 2000 m s^{-1}, considerably higher than the speed of sound in air, 330 m s^{-1}. The resulting high-pressure pulse of air is pushed outwards by the hot explosive products. In order to allow comparison between charges of different sizes and compositions, explosive engineers conducted experiments using the simplest scenario possible: a bare explosive charge in an empty, flat field. This has resulted in a large body of blast wave literature based around the classic 'Friedlander' blast wave form [13].

Figure 5 shows the pressure time profile of this classic blast wave. After the initial rapid rise of the blast wave there is a region of positive pressure, accelerating outwards from the explosion. The speed of the gas moving behind the blast front, in the so-called 'blast wind', can be as high as 500 m s^{-1}. This is followed by a release wave that drops the pressure below atmospheric pressure. The release occurs as both air and product gases have expanded outwards, away from the place where the explosion initiated. As a result, the explosive products have expanded and therefore performed 'work' on the surroundings and started to cool. This leaves a partial vacuum in the region of initial explosion and this lower pressure now causes air and gases to flow backwards over a longer period to equalise the pressure. The resulting push–pull movement experienced in the blast wave can be especially damaging to structures and humans.

This simple waveform was often observed in much early research into the effects of blast on humans [14, 15]. These studies used data either from open explosive ranges or with well controlled blast reflections from single walls or barriers. It is important to note that in a cluttered urban or vehicle environment, a casualty will experience a number of waves: those directly arising from the charge and those reflected from a wide variety of surfaces, arriving from different directions. In general the amount of reflected pressure waves can be equated to more damage and injury.

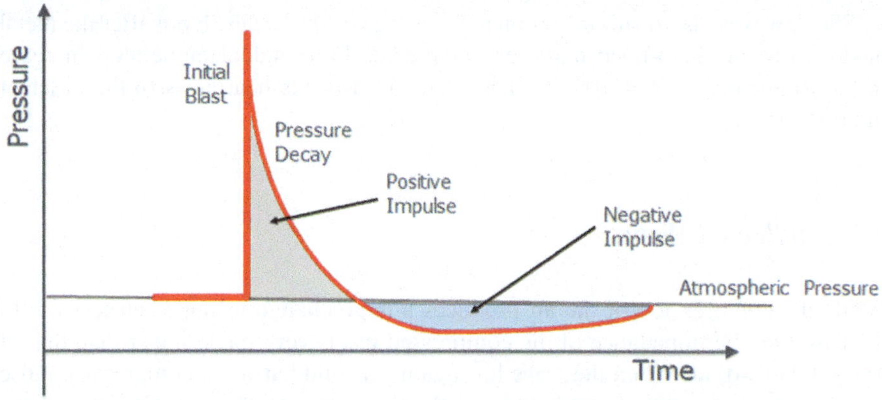

Fig. 5 The classic Friedlander form of a blast wave. Redrawn by W.G. Proud

2.4 *Change in Mechanical Impedance of Air in a Blast Wave*

In the discussion above the energy deposited in a material can manifest itself in a number of ways. It can give the material kinetic energy, for instance, and it can compress the material increasing its temperature.

In many engineering applications, solids and liquid are often assumed to be incompressible. In shock wave studies this is not the case and compressions which halve the volume occupied by a material are relatively common. Gases are by comparison, very compressible. This means that blast waves become supersonic at relatively low pressures, and are also associated with a very large change in the mechanical impedance. Mechanical impedance is a material property determined by multiplying the material sound speed by the density. It is a measure of the stress flux (impulse) delivered by a material and is used to calculate the amount of a shock of blast wave energy that is transmitted from one medium into another. If there is a large difference in impedance between two materials (e.g. air and a concrete wall), then most of the energy will be reflected from the surface [16].

However, a major difference between shocks in solids and liquids, and blasts, is that blast waves from munitions tend to have durations measured in milliseconds, while shock waves in solids exist on timescales of microseconds. This thousand-fold difference in duration is why the relatively modest pressures seen in blast waves can produce more movement and damage than the much higher pressures in shock waves, a greater overall impulse is delivered. Greater detail can be found in Kinney and Graham [17], Needham [18].

Also, a large change in impedance implies significant compression and this is often associated with an increase in temperature. This can result in burns and combustion. Inside a vehicle, the heat deposited into a material can be divided almost evenly between the heat from the intense light flash associated with the detonation or gas compression and heat from physical contact with the hot gas. The timescale for this energy transfer is of the order of milliseconds.

Shock waves also result in significant heating within metals. Even after the metal has been dropped back to normal stresses there may be a residual temperature increase in the fragments of over 100 °C. The timescale for this heating is in the order of microseconds.

2.5 *Reflected Waves*

While the compression of the air produces a large change in impedance, it is still the case that the impedance of the compressed gas is very much lower than that of any solid or liquid. When the wave hits against a solid barrier, a compressive pulse is transmitted into the object but also reflected back into the gas. Calculating the properties of the reflected wave indicates that the stress level almost doubles. So proximity to barriers can easily increase the effects of an explosion.

In the case of an explosion in an enclosed space, or in the partially confined space below a vehicle, there is time for multiple wave reflections; this leads to complex blast waveforms of widely differing stress histories [13]. Within a vehicle the reflections add to the injuries seen, over and above the push–pull effect seen in the relatively simple Friedlander waveform (Fig. 6).

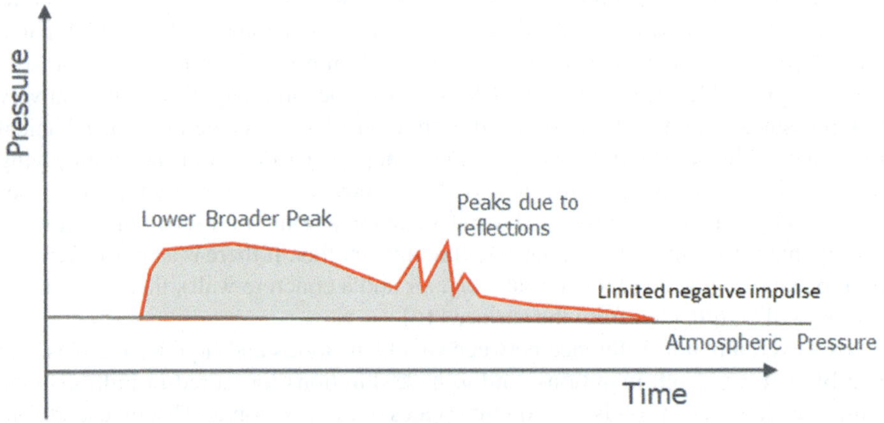

Fig. 6 Waveform showing a possible form of loading seen due to reflections internal to a vehicle. Redrawn by W.G. Proud

2.6 Comparing Explosives Scenarios: Scaled Distance and TNT Equivalence

Munitions and explosive charges come in a range of sizes and vary in materials. In the technical literature, the term 'scaled distance' is often used to relate the effects of large explosive charges over tens of metres to those of small charges at close range. This is useful as it allows the effects of small-scale experiments to be extended to larger explosives charges.

The two terms used in virtually all scaling equations are (i) the explosive mass and (ii) the mathematical cube of the distance between the charge and the target, i.e. the increase in volume through which the energy is dispersed [17].

Often other factors are included. One of the easiest to conceptualise is the distance of the charge above the ground. If the charge is in mid-air, the energy expands evenly in all directions, effectively spreading the energy through a sphere. If the charge is near the ground, the effects of wave reflection occur, and the energy is concentrated into a reflected wave from the ground of a roughly hemispherical shape. Within the hemi-sphere the energy, pressure etc. is more or less doubled compared to that of the mid-air charge. This reflected wave is also why a mushroom shaped cloud is sometimes seen above explosions.

Another simple comparative scale between explosive types is that of 'TNT equivalence'. Historically TNT was a widely used material, which could be easily melted and cast into a variety of shapes, unlike many other explosive materials. Given its castable nature many experiments were conducted, resulting in large databases. TNT equivalence is a simple factor that allows a well-defined reference point for the broad comparison of the effects of a non-TNT charge to be estimated.

2.7 Vapour Cloud Explosions

With the use of flammable gases such as methane, hydrogen, and acetylene in domestic and industrial settings, accidental leakage can result in build-up of reactive vapour clouds. The over pressure from unconfined gas clouds of this type cause little to no damage to structures. It must be noted here that some weapon systems use a fuel–air effect, in this case it is a mix of droplets of liquid fuel mixing in air resulting in high energy output over a much longer time.

Vapour cloud explosions (VCEs) are generally ignited by a spark or flame, causing a deflagration (rapid burning) or detonation (associated with a reactive shock wave). Which pathway is dependent on a number of factors. When a gas cloud is not mixed with air and its concentration is above its lower flammability limit (LFL), it will deflagrate. When the gas and air are mixed, ignition and burning is much quicker, leading to a rapid expansion forming blast waves. There are three main parameters that determine the overpressure from a VCE: burning velocity, degree of confinement, and cloud geometry. Burning velocity is how quickly the flame spreads in the unburnt cloud from the ignition point. The geometry of the cloud describes the volume where the flame can spread, i.e. the shape of the pipe, hallway or stratification. The degree of confinement is the area of solid boundaries where gas expansion cannot occur, hence causing expansion of products to more vented areas. VCEs are much more common in industry than in dwellings. However as discussed in 1.2, gas leaks do occur in homes occasionally resulting in injuries and fatalities. VCEs in confined spaces are very sensitive to obstacles and the amount of venting available. A VCE is usually less destructive than one caused by a detonation since there is no supersonic blast to cause substantial damage to structures. When a hydrocarbon-air VCE occurs in an open space, the flame front moves in a laminar fashion at a subsonic velocity relative to surrounding gas in the order of 3–4 m s^{-1} and reaches maximum velocities of 20–25 m s^{-1}.

In an environment containing obstacles, as the flame front moves around those obstacles turbulence is created. The amount of turbulence caused by objects in a confined space can result in flame fronts jumping ahead of the leading laminar front, increasing the rate of combustion. This acceleration results in a turbulent deflagration and the increased front velocities can cause a deflagration to detonation transition (DDT), resulting in a supersonic blast front shocking the unburnt gas in front [19].

The strength of a blast from a vapour cloud is strongly dependent on its fuel–air stoichiometry. If all the fuel and air in a mixture is consumed, a stoichiometric mixture, this will result in the most violent reaction. The location of the ignition is important, ignition at the outer edge of a gas cloud generally produces a turbulent acceleration, increasing the energy release rate while ignition at the centre of a cloud allows less distance and time for such effects.

3 Human Injury

Wound ballistics address the effect of impact and rapid accelerations on the human body, and processes which occur on a timescale from microseconds up to a few seconds. There are two main schools of research in this area. The first comes from the physical sciences and engineering, which tend to concentrate on the transient loading. The second comes from the medical profession who deal with the longer-term consequences.

The injury process tends to be dominated by two effects. Firstly, the blast is directional, the wave strikes one part of the body first, compressing it into the so-far unaffected parts of the body. This is the result of higher sound speeds in the liquids and solids of the body compared to the blast wave in air. There is differential compression as softer biological organs push against harder ones. The second effect, occurring milliseconds later, when the body is completely enveloped by the blast, is a general compression. The internal structure of the human body means that the compression of a particular body part during the blast is difficult to precisely determine, but compression is more pronounced in air-filled organs.

The two organs most affected by blast waves, those with the greatest air content, are the lungs and the ears [14, 15]. Short pulses are more survivable as there is less time for compression to occur so overall the lungs are compressed less. However, the lungs have a certain volume and once fully compressed, little more damage can be done and so the survivability curves tend to flatten off.

It is of interest to note that the survival rates are much worse if the subject is close to a wall or other large structure. In this case, the blast wave reflects from the wall and subjects the body to a second higher pressure pulse. As a general approximation the pressure seen from waves reflected from walls and the earth tend to be slightly less than twice that of the initial pulse.

The effect of a blast wave moving along the long axis of the human body; here the human body is stronger as the spinal column will support the ribcage in resisting compression. The chances of survival are much higher, and approximately nine times more pressure is needed to injure the subject than if they were next to a wall.

Another effect seen over a number of experimental studies using a variety of animals showed that creatures with small lung volumes per unit body mass were more susceptible to blast; also those animals with denser lung material were more vulnerable. The small lung volume can be thought of as reducing rapidly over a short period of time, while larger lungs take longer to collapse and so may be only compressed to a small extent during the compressive phase of the blast wave. The second point is that denser lungs are better impedance matched to the surrounding tissues, so more stress is transmitted into the lungs by the surrounding rib cage. Physiologically, the condition of blast lung is well known in the medical community, where some patients who are apparently well initially will die within 24–48 h.

Damage to the hearing is mainly by failure of the ear drum, caused by the pressure difference between the air within the skull/ear canal and the outer atmosphere. The blast wave pushes the ear drum membrane inwards, tearing it. Depending on the blast wave pressure the membrane can fail from a simple rupture (level 1) which can repair itself given time, or a catastrophic rupture, that shreds the membrane (level 2). Finally, at high levels of pressure the supporting structure around the ear is torn away completely (level 3) [6].

3.1 Non-penetrative and Acceleration Injuries

These types of injury are most likely to occur for occupants of vehicles, passing close to or over an explosive device. Civilian medical knowledge of these injury types are well-known due to the high number of traffic and pedestrian-car accidents seen each year. Aerospace studies also provide significant data. Table 3 compiles some values of the acceleration required to incapacitate a subject [20]. The variation in the units used is due to differences in methodology and decisions based around the mass of the head, body etc. In the area of blast injury in vehicles, much effort has focussed on lower limb injury, especially the phenomena of 'deck slap', where the floor violently moves upwards, violently accelerating the feet and legs of those standing on the floor. This type of injury was first noted on metal warships.

Often impacts will not penetrate the skin but will cause pain and internal injury. Most of the studies from research centres such as the US Army land warfare laboratory use energy per unit area to compare slow-moving, large projectiles with smaller high-velocity projectiles. The US Marine Corps states that energy delivery of 6 J cm^{-2} will produce "dissuasive" pain [21]. Non-penetrating impacts with a total striking energy between 40 and 120 J can cause dangerous injuries while 120 + J can easily cause death.

Effects seen in blunt impact include effects due to differential acceleration or stress wave overlap. The "two strike" phenomenon is an example of the first and shockwave focussing of the latter. The two-strike process is seen when a hard projectile strikes a target of finite thickness. The target material initially moves with the projectile but, as stress waves reflect from the rear of the impacted material, the target loses contact with the impactor for a period of time and is then struck again. For the human body, the separation between these impacts may be up to one second [22]. Measurements in gelatine have shown that while the stress in the second pulse is lower in height, it is of longer duration; overall this can transfer as much energy as the initial impact.

The effect of shock-focussing is based on the shape of the target [22]. Overlapping release waves can be particularly relevant to internal injuries. For projectiles, initial impact sends a compressive pulse into the material. However, when the release waves originating from the sides of the impact zone overlap along the central axis of the impact zone, a tensile region is formed where multiple cavities-injuries can be nucleated. This is referred to as a cavitation and is a widely studied phenomenon in liquid transport and ship-propeller design [22].

Whilst some burn injury is caused by direct contact with hot gasses (from convective burning, and hot fragments), it is estimated that up to 50% of the heat energy transfer can come from radiation, the intense flash of light seen in the detonation process [23].

Skin is thinnest on the eyelids (~0.5 mm) and thickest on the back (5 mm), with an average value of 2.00 mm widely used. When skin reaches a temperature of 58 °C pain is triggered. Superficial burns affect the outer layer of skin (0–0.1 mm deep), causing reddening. Partial thickness burns involve the destruction of the outer skin

Table 3 Accelerative forces and durations required for complete incapacitation

Body part	Acceleration/impulse	Duration
Head	150 g* or more	2 ms or longer
Pelvis (vertical motion)	23 g or more	7 ms or longer
Spine (instantaneous)	2100 + N m	<1 ms
Spine (longer loading pulse)	1160 + N m	30 ms or longer

Note 1 g = 9.82 m s^{-2}

cells (0.1–0.5 mm deep), causing blistering: new skin will grow out from the hair follicles. Extensive partial thickness burns can kill due to fluid loss infection. Deep dermal burns (0.5–2.0 mm deep) destroy the hair follicles and therefore skin will not regenerate: skin grafts are required to avoid long term scarring.

Energy absorption in the skin is not a simple process; reflectivity and light absorption vary with wavelength. For high temperature sources (4000 K), white skin will reflect 40% of the incident visible light. The 60% of energy absorbed can penetrate relatively deeply and is absorbed through a thickness of 2.0 mm. Black skin reflects less and absorbs strongly in the melanin containing layer, concentrating energy in a smaller volume. This means that white skin is more resistant to thermal heat flashes. It must also be added that white skin performs poorly compared to black skin in relation to UV-induced skin cancer for the same reason: the depth of light penetration.

Results using black-body sources at 2000 and 4000 K have shown that 4000 K radiation is less damaging than 2000 K radiation because the wavelengths from the 4000 K source fall in the visible region and are strongly reflected [12, 23]. The cooler 2000 K source emits greater energy in the infra-red, where the skin absorbs more. Simple ad-hoc solutions like sun-creams have been used to defeat some of the radiative effects.

3.2 *Penetrative Injuries*

In principle, the penetrative process is very simple: the strength of the skin is overcome and the projectile pushes through the bulk of the body, delivering kinetic energy to the surrounding flesh. If a sufficiently high velocity is used, the projectile will perforate the body and an exit wound will be formed.

However, because of the complex internal structure of the body, differentiation of organs and cell types, bones, etc., this is never quite so simple. Data sources commonly used include crime scenes, military operations, and laboratory studies. Of these, police related publications tend to deal with aspects of legal process: was a shooting legitimate, was the use of a particular ammunition type justified? The military tend to be dominated by tactical considerations, effective range of weapon types, and body armour. Laboratory studies tend to use animal cadavers and inert stimulants in the main, with a small number of human cadaver studies being performed.

To put the levels of energy and pressure in perspective, a standard NATO round of 7.62 mm diameter has a muzzle velocity of ~830 m s^{-1} and a kinetic energy of ~1,000 J. The impact pressure of such a projectile will be up to 320 kbars if it was a steel-on-steel impact, though this pressure is much lower, ~15 kbars if impacting water, gel, or the human body.

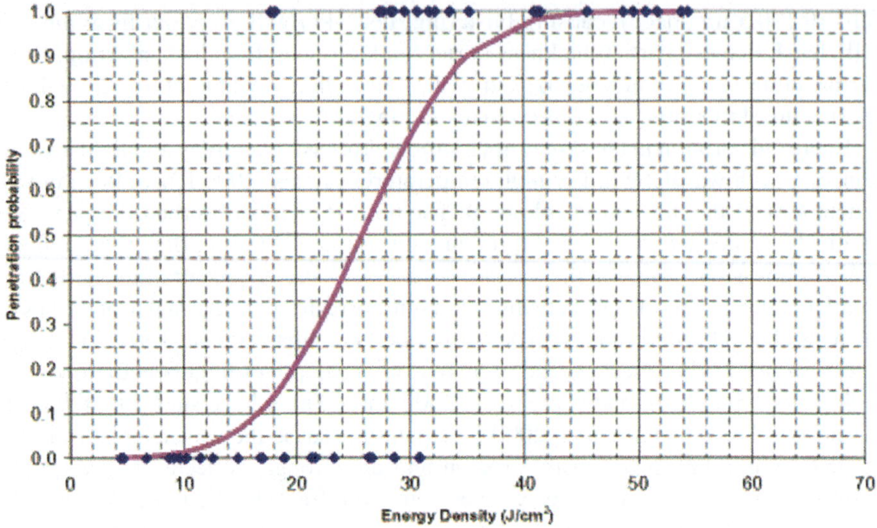

Fig. 7 Probability of skin penetration against impact energy. Figure reprinted from the BALLISTICS 2008: 24th International Symposium on Ballistics, 2008. Lancaster, PA: DEStech Publications, Inc.

3.2.1 Skin Penetration

While skin penetration experiments tend to be assessed on a yes–no basis, Drapela et al. [21] suggested for a 50% probability of skin penetration an energy level of 25 J cm^{-2} (Fig. 7). This is based on statistical averages of hit locations on the body taken from a wide database of injuries. However, they suggest the use of a lower value of 10 J cm^{-2} for practical purposes, to allow for the probability of impact on less robust areas of the body, such as the face, hands etc.

3.2.2 Pencilling

Following police reports there have been a number of incidents of victims wearing Kevlar and other protective clothing being injured or killed by gunfire. This is called "pencilling" and has been studied by researchers at the Defence Academy of the UK [4]. The basic process is easy to understand. The protective garments are either worn loose fitting or are not backed by a solid support; military personnel would tend to have the garment pulled as tight as possible, while allowing the wearer to move or to be backed by a support of some kind. The bullet strikes the outer surface pushing both itself and part of the garment into the body. The garment provides a means by which the load is spread over a wider part of the body, but no penetration of the garment occurs. Afterwards the garment can be pulled from the wound leaving a mildly distorted projectile in its folds.

3.2.3 Projectile and Fragment Injuries

Projectiles can be defined as a shaped components designed to fly over some distance, while fragments are the shards of metal, or other material, from the casing of a shell. Shrapnel is a form of projectile, shaped pieces of material placed within and around an explosive charge with the express intention of them being thrown violently by the blast: the invention of the British Army officer Lieutenant Henry Shrapnel.

In the initial stages of impact, the projectiles send out a compressive pulse into the material. For a small projectile, the release waves originate from the sides of the impact zone. When these waves overlap, a tensile region is formed where tears and cavities can be nucleated.

This effect happens in the body with cavitation occurring in front of the projectile, tearing and weakening the structure. This is coupled to two more effects. Firstly, the projectile pushes the target forward and also pushes the material out laterally. This results in a radial expansion of the hole carved out by the projectile, thus widening the overall cavity. It is this momentum transfer which opens the cavity, not the popular misconception that the cavity "inflates" by gas rushing in. This would only be the case if the gun was at very close range, e.g. significantly below 1 m.

The second effect occurs if the projectile is moving faster than the speed of sound in the air (~332 m s^{-1}). In this case a low, sub-atmospheric pressure region with turbulent flow is found in the zone immediately behind the projectile. The effect of this is that air and other debris can be sucked into the expanding wound cavity.

This initial cavity is not permanent as the flesh will spring back, forcing the air out of the cavity and reducing its volume [24]. The maximum volume is referred to as the temporary cavity volume; it can easily be in excess of 200 cm^3 for a small arm round of 5.56 mm diameter, travelling at 800 m s^{-1}. Figure 8 a shows the large extent of cavitation seen in simple gel targets. The edges of the temporary cavity are not smooth and in the body the varying nature of the body organs, bones etc., result in significant tearing between material types [25]. Figure 8b shows how a small cavity can deflect the projectile path.

3.2.4 Kinetic Energy and Projectile Flight Path

In the broadest terms, the extent of injury is dependent on the amount of energy transferred to the target. In many cases a low velocity projectile which penetrates but then stops inside the body will produce more overall damage, and probably push the casualty to the ground, than a higher velocity projectile which enters and exits with only a small reduction in velocity.

The penetration depth and energy loss of 7.75 mm pistol rounds at 300 m s^{-1} into gelatine showed a penetration depth of the order 65 cm for 10% gelatine, 36 cm for 20% gelatine, energy loss rates averaged 3.48 J cm^{-1} and 5.93 J cm^{-1} respectively [25]. Pig flesh, generally regarded as a good human simulant, showed a loss rate of 6.90 J cm^{-1}.

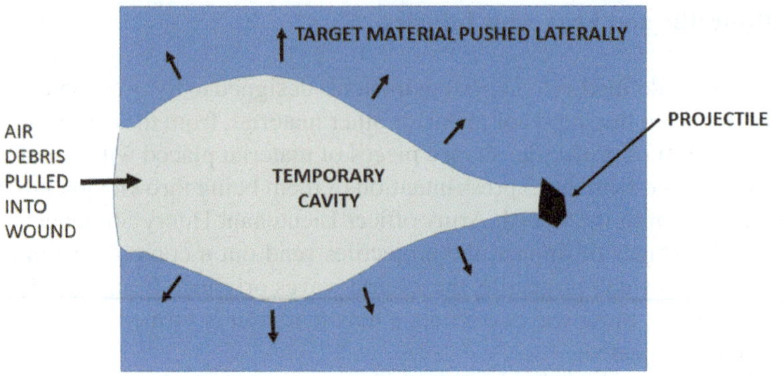

(a)

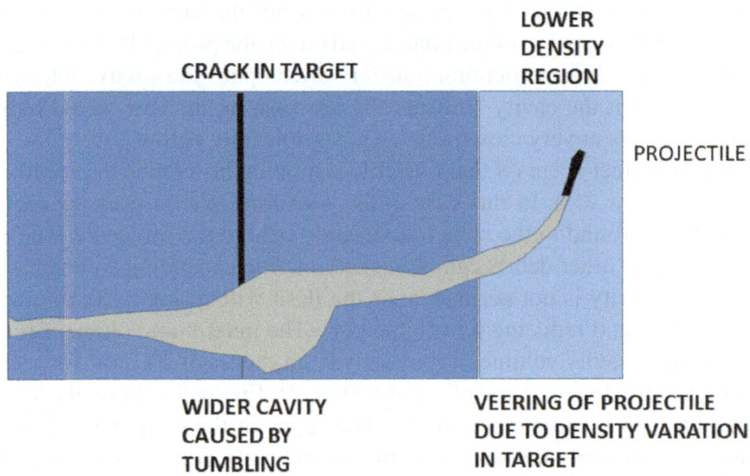

(b)

Fig. 8 **a** Cavities in gelatine caused by penetrative impact. **b** deviation of a projectile by a small gap in the gelatine. *Credit* W.G. Proud

The path of the projectile can vary markedly even in homogeneous materials like soap. Assumptions of straight-line motion can be wildly inaccurate. The initial angle of impact, organ structure, presence of bones, lungs etc., can all cause the bullet to change direction by 90° or more in a relatively short distance. Bones do not tend to stop bullets, except if the bullet enters at a very low speed (<100 m s^{-1} or much lower). Bones tend to shatter while deflecting the bullet, causing bone splinters which can cause extensive secondary damage.

A 1995 Danish study [26] examined fragmentation of 7.62 mm NATO systems. Rounds from one manufacturer were found to be unusually susceptible to shattering, thought to be caused by the impact on the human body. Initially, evidence was gathered from crimes, accidental shootings etc. However, controlled studies showed that this round would fragment after travelling ~100 m in free flight, at which point it had also been significantly slowed by the air. Fragmenting or explosive bullets, dum-dums etc., are subject to prohibitive legislation dating from the Hague convention of 1899. In practice a projectile may fragment in an unexpected fashion due to manufacturing processes or by ricochet or by penetrating through an armour system.

3.2.5 Ocular Damage

The eyes are particularly susceptible to impact damage. Some of this research has been driven by the occurrence of paintball injuries seen amongst children left alone with paint guns and not wearing any goggles. The energy requirement for significant damage is reported as 2.5 J cm^{-2}. However, a recent study using pig eyes impacted by paint-balls has shown the required levels for different degrees of damage [27], Fig. 9. It should be added that all these levels of damage are quite serious and would require significant medical intervention.

The damage mechanism tends to be compressive, with failure of the joining materials between parts of the eye. However, rapid rotation of the eyeball can produce the shearing of the optic nerve. The rupture of the eyes is caused by compression of the eye causing a high level of pressure inside the central cavity, the liquid of which ruptures then pushes through the outer shell of the eye by the tensile (hoop stress, around the circumference) failure of the shell, but generally this extreme level of injury requires more effort than penetrating the skin.

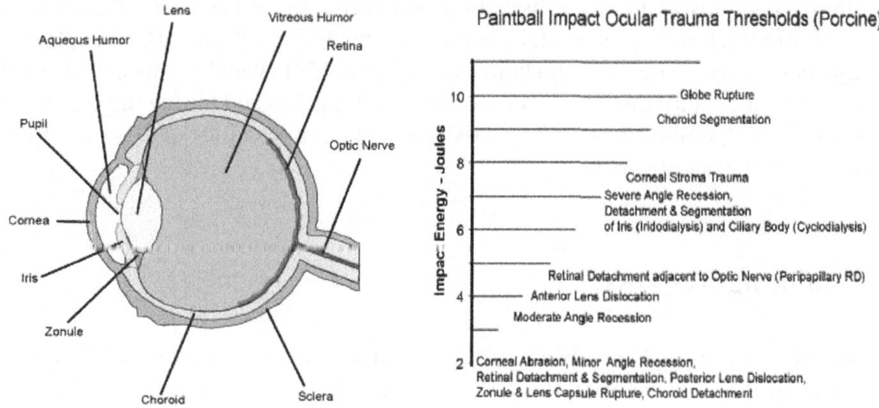

Fig. 9 Structure of the eye and injury-energy level as determined by impact on pig eyes Figure reprinted from the BALLISTICS 2008: 24th International Symposium on Ballistics, 2008. Lancaster, PA: DEStech Publications, Inc.

4 Vehicle Response

The use of improvised explosive devices in both warzones and in terrorism has been a hallmark of the late 20th and early twenty-first centuries, with trains and road vehicles forming targets of increasing importance.

From a blast science point of view, an explosion within a vehicle is very similar to that of a very weak building. The presence of windows and doors in the vehicle allow a rapid venting of the blast wave and explosive products. A review of the casualties in the 7/7 bombings in London, where buses and underground trains were attacked, will show that the maximum number of casualties was produced on the underground train in a narrow tunnel: the tunnel walls providing a strong confinement on the explosion, so magnifying its effect.

The explosion of IEDs under vehicles is a special case, however, as the vehicle tends to lift with the blast wave, so protecting the occupants from fragments and the heat of the products. Within vehicles, the injury patterns tend to be dominated by leg injuries caused by the violent acceleration of the floor or head/neck injuries due to the occupant being accelerated against the vehicle roof. The acceleration of the floor produces significant lower limb injury to the knees, ankles, and hip joints. This has been the focus of much recent effort; comparisons with data from civilian traffic accidents, while having some relevance, is also not a good model. Blast injuries tend to have more intense local damage but also involve significant injuries spread over the body, i.e. it is very violent compared to a high-speed road accident.

The effect of buried charges effectively replaces the effect of hot, high-pressure gas with that of compacted soil and stones being violently thrown against the vehicle. Overall, the impulse delivered is higher for buried charges. For example, the velocity of a 25 mm thick high-grade steel plate loaded by either a flush-surface explosive charge or a buried charge (50 mm below the surface) [28], would produce loads well in excess of the injury limit. An average person falling 3 m and hitting the ground at around 8 m s^{-1} would run significant risk of injury. At velocities of 20+ m s^{-1} the effect would be catastrophic.

5 Building Response

While blast waves and the ensuing fragmentation can result in severe injury and death, in an urban environment the collapse of buildings can lead to more widespread damage and injury. Depending on the position and size of the charge used, structures will experience a combination of the blast wave and a stress pulse due to forces transmitted through the ground, i.e. a mini earthquake of short duration [29, 30].

As mentioned earlier, the blast energy transferred to the structure depends on the impedance difference between the blast wave and the local structure. When impinging on a solid object, blast waves will both reflect from and diffract around it. The reflected wave will then form a region of further compression of the air, producing the reflected pressure that exerts on the structure. The reflection could be categorised into two genres: regular reflection and Mach reflection. Regular reflection happens if the incident angle is small and the wave encounters the surface almost perpendicularly or <~40°. Above this threshold, the reflected wave along with the incident wave will produce a Mach stem. A Mach stem can be thought of as a fast-moving pressure front of double the pressure of the initial wave and so increased damage will be found in the areas affected by such a phenomena.

Generally, the largest response in structures is caused by blasts with fast rise times, e.g. high explosive detonations. However, the response of a structure is linked strongly to the ratio of loading duration to the natural resonant period of the structure. There are 4 response categories to the shape of a blast wave. Cullis [32] defines them as.

Category A: Impulsive loading where the duration of the wave is short with respect to a structure's oscillation period. This means that some of the load is absorbed by the inertia of the structure reducing the amount of deformation. This is often the case where high explosives are involved.

Category B: A quasi-static load which has a long rise time, and its duration is long compared to the oscillatory period of the structure. The load is equivalent to that of a static load. This is more common for gas explosions then high explosive events.

Category C: Again, a quasi-static load, but with a short rise time and long duration compared to the structure's oscillation period. This category could cause a structure to experience more deformation than that caused by a static load.

Category D: A dynamic load which is a combination of an impulsive and quasi-static load. The blast wave duration is similar to the period of oscillation for a structure. If the blast wave rise time is around half its duration this may produce a greater response than one caused by a static load. Such long duration, high impulse events are more associated with explosions with significant 'heave' e.g. ANFO.

For a blast inside a building, the physical process of reflection, potential for Mach Stems, is the same as external explosions, however, the presence of multiple walls, doors, and furniture will produce more complex pressure pulses due to reverberations and partial confinement. Overall, the interaction will always depend on the geometry, weight, scale, material, total area of the object exposed, and the surroundings [29].

5.1 Pressures and Impulses: Materials and Structures

If a strong blast loading is experienced, the building may start to move. The regimes of impulsive, dynamic, and quasi-static loading can be visualised by the use of pressure-impulse curves, Fig. 10. Different regions of the curve represent the types of loading, categories A, B, C or D as previously mentioned.

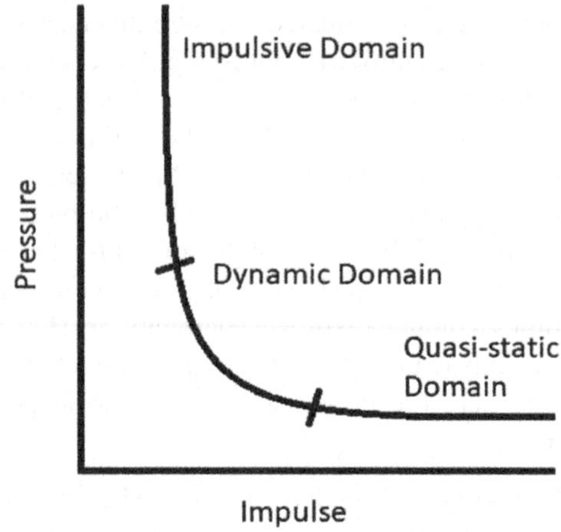

Fig. 10 Simple pressure against impulse diagram showing the impulsive, dynamic and quasi-static loading domains of a blast. *Credit* J. Cooper

The limits for pressure and impulse can be defined where the minimum pressure or duration, for a certain amount of damage to occur, for impulsive loads and quasi-static loads respectively [33]. For high explosives the response close to the charge will be in the impulsive zone, however, with increasing distance the pressure and impulse will drop. Between the impulsive and quasi-static region in Fig. 10 is a mixed zone. This mixed zone is the type of loading seen in blasts caused by gas explosions. Jarrett in 1968 empirically created pressure versus impulse diagrams based on damage sustained to typical homes and factories in Britain during World War II [34]. Such diagrams also exist for structural sub-components and the levels of damage for impulse-pressure loadings for a reinforced concrete (RC) wall [35].

5.2 *Progressive Collapse*

If sufficient sub-components break, a situation of progressive collapse (PC) begins: local failure which progresses to adjacent elements, causing a relatively slow moving but extensive collapse. PC can be classified into commonly occurring types [36–41]:

(a) Pancake
(b) Domino
(c) Zipper
(d) Section
(e) Instability
(f) Mixed

The detailed dynamic structural response is complex if all the factors such as structural shape, nonlinear material behaviours, and time-dependent deformations are considered. The problem is often simplified by making assumptions: Elastic Single Degree Of Freedom (SODF) models and Elasto-Plastic SDOF models are most commonly used. Both substitute the real structure with a system of idealised connectors, each consisting of one concentrated mass and one weightless spring. The motion of the building and its resonant frequencies, global and local, can be calculated. If a failure strain is defined for each spring, then the likely collapse points in the structure can be defined. Such models are widely used to predict the response of structures to storms, earthquakes, and other loading criteria of which explosive loading is one option. Elastic SDOF models are suitable for the elastic region, where after loading the structure will return to its original shape. A more complex approach is to use Elasto-Plastic SDOF models, which are appropriate for systems where the plastic region is important. Structures with extensive metal skeletons generally fall into this category.

Local structure response to blast loading can be defined as flexural motion or localised shear, brought about by an explosion taking place close to the structural elements, the outcome being determined by the distance from explosion to the target and the ductility of the structural element. Global response is the consequence of long-time exposure to transverse loads. It possesses two failure limits: membrane/bending failure and shear failure.

5.3 *Reinforced Concrete*

Reinforced concrete is a key component in the construction of both houses and high-rise buildings. Its ability to support heavy weights and to be poured on-site makes it a popular and essential building material. Its high strength is due to steel rod reinforcement, enabling it to withstand high tensile strains. Reinforced concrete (RC) is usually subject to 'pre-stressing' to increase the overall strength, maximum tensile strain and ductility [42]. Since reinforced concrete is common in construction of dwellings, its behaviour under blast loading has been widely studied.

However, one issue is how to deal with older building stock. Ronan Point was a block of high-rise flats constructed in the late 60s using the Larsen-Nielson system of prefabricated concrete structural elements. On May 1986, a gas explosion on the 18th floor lead to the progressive collapse of the entire southeast corner of the building. Pearson et al. [43] produced a report analysing the collapse. The explosion was found not to be particularly significant as objects taken from the apartment of the where the explosion happened indicated they had been subject to pressures <70 kPa. The Building Research Station and Imperial College London performed experiments to determine how much force Ronan Point could withstand. The internal walls were discovered to be displaced by a small pressure of 19.3 kPa [44]. It was also estimated that the internal walls could be displaced by a pressure of 1.7 kPa and the exterior walls could be displaced by 21 kPa [45].

It was found that the structure was fundamentally flawed due to poor construction practice. When the blast occurred the corner walls of the apartment on the 18th floor were pushed outwards. These walls formed the only support for the floor above leading to the progressive collapse up to the 22nd floor. The combined weight of the collapsed structures caused the wall supporting the 18th floor to fail. This lead to a domino effect resulting in the collapse of the entire southeast corner. Four people died in the building and seventeen were injured (one later dying in hospital). The deaths were confined to the floors above the 18th floor. These deaths were due to structural failure which brought about stricter building codes and minimum requirements for building materials [43].

Such incidents, though rare, occur globally. In Romania, a flat on the 4th floor of a 5-story apartment building was completely destroyed by a VCE explosion. The explosion was caused by a leaking propane cylinder ignited by a light switch. The blast also impacted surrounding structural elements with various levels of severity. The damage to the flats was investigated [46] and a structural analysis was carried out to determine the loads exerted on elements of the structure. The apartment building was constructed with 30 cm RC walls and concrete facing slabs with 15 cm interior panels. Floors in the building were constructed from 14 cm RC. In the flat where the blast occurred, the concrete wall panels in the flat were completely destroyed. The blast also displaced the concrete facing slab by 55 mm and the RC upper and bottom floor experienced deflection of up to 300 mm. All windows and doors were broken. The surrounding structural members of the building were also damaged. The concrete facing slabs above and below the 4th floor were displaced by 10 mm and the RC floor of the above the neighbour's apartment had deflected by up to 50 mm. The walls and floors of the flats crossing above or below the blast site were cracked due to dynamic effects of the shock.

When a blast wave interacts with concrete or an RC column, a wave is transmitted within the material. This is reflected off the free surface at the opposite side of the blast loaded surface and can cause the concrete to spall. A concrete/RC column or wall will spall if the net primary stresses exceed its tensile strength. If the stresses causing spall are strong enough, it can lead to fragments flying from the concrete with high velocities. This poses a serious threat to personnel both within and outside the building [47]. Building experiments by Wang et al. [48] on RC slabs found that close in spall damage is insensitive to column stiffness and boundary conditions but is limited to a local area. The amount of damage depended heavily on standoff distance and charge weight. Li and Hao [47] found that by increasing the level of reinforcement, increasing column depth and changing the layout of reinforcement decreased the levels of spall. Spalling is much less likely to cause structural failure than flexure and shear but is still important due to possible fragmentation of RC.

For a VCE that does not detonate, it is important to note that the rise time to the peak pressure would not be as sharp as the ones used in these RC studies. The effects can be analysed with P-I curves using the increased blast duration of the deflagration. The only caveat to this argument would be that high explosives are, generally, concentrated into localized volumes, while VCEs tend to occupy much larger and widely dispersed volumes. Similarly, the effects of confinement on the RC columns and slabs that have been modelled and tested were not in confined spaces, confinement will increase the duration and intensity of loading. The high durations means RC structures experience pressures associated with blast waves as a quasi-static load, particularly in the case of deflagrations.

5.4 Steel Columns

Another common building material in modern building design is steel. It is used to form the skeleton of many high-rise buildings and is often associated with concrete panels in older buildings and glass panels in newer constructions. So it is vital to evaluate the damage thresholds under blast loading. RC is much better at resisting the close in effects of impulsive blasts than steel, which is much more suited to resist quasi-static loads of lower pressures [17]. The response of RC to blast is relatively predictable as it is used in large sections, whereas steel is used as a skeleton to many structures as smaller elements. This means steel components are slender compared to RC components, leading to local instabilities and unpredictable ultimate capacities [49].

Assessment of stress levels under dynamic loading is challenging as many stress combinations may occur, making modelling difficult. To show the level of complexity involved, the research of Dragos and Wu [50] is presented in some depth. A finite element model (FEM) and single-degree-of-freedom SDOF approaches were used to model the response of steel columns from blast waves under various axial loads. The model considered instabilities within the column due to axial load effects. From the model, pressure impulse (P-I) diagrams were produced showing the effect of axial load. The columns [50] used in the study had a yield strength of 470 MPa and a Young's modulus of 180 GPa. The columns both had the same dimensions with angles 102 mm wide and 10:3 mm thick. The web, i.e. centre section of the column, was 6.6 mm thick and the total depth of the column was 160 mm. One column was simply supported and 4 m long, while the other was 8 m with fixed supports. An axial load of 1480 kN which resulted in buckling was determined to be the critical load. The axial loads used in the models were 0, 175, 350, 500, and 700 kN. A numerical finite element model (FEM) was carried out which produced P-I curves for both specimens. Validation of the SDOF approach was made by the FEM as various blast load durations could be carried out. P-I diagrams for both the 4 and 8 m span columns were produced and it was found that the longer span columns are more susceptible to failure for all axial loads [50].

5.5 Masonry Walls

Masonry walls are widely used and many of the buildings in the UK are constructed from bricks and mortar. Many old structures were also constructed in this way and over time the materials have embrittled, leaving them vulnerable to blast loading. Furthermore, the fracture of internal non-load bearing and load-bearing walls can lead to high-speed debris which can cause further damage.

Wei and Stewart [51] studied the blast response of unreinforced brick masonry walls to blast from high explosives, however, the results can be applied to VCE detonation events. They concluded that mortar and brick strength has an insignificant effect on a wall's response to large blast loads. Under low blast loads, the mortar and brick strength had a large effect on maximum wall deflection. Wall thickness and the boundary conditions, i.e. how the walls connected with the rest of the structure, had a significant impact on the amount of damage.

Li et al. [52] investigated the effects of vented gas explosions on unreinforced clay brick masonry. The study used a RC chamber with internal dimensions of 3 m × 2 m × 2 m with one masonry brick wall. A vent window was used to moderate the pressures in the chamber. The bricks were 240 mm × 115 mm × 90 mm, constructed with English or Running bonding patterns. The gas used for the VCEs was methane and produced deflagrations. The VCE blast pressure was to a great extent uniform in the chamber because the pressure wave velocity is much faster than the flame front and the natural frequency of each wall is shorter than the load duration. The gas explosions can be likened to quasi-static loads. Increasing thickness of the wall also caused the peak wall displacement to change: the 240 mm wall was displaced significantly less than the 115 mm wall. At 20 kPa, the 115 mm wall collapsed while the 240 mm wall was still reusable. The boundary constraints of the wall also influenced the response: one-way walls where the right and left sides were isolated from the RC chamber by plastic film, and two-way walls that were bolted onto the RC chamber with steel bars on the right and left sides of the walls. It was seen that a two-way wall under the same test conditions experienced a peak displacement 95% smaller than the one-way wall.

Analysis on autoclaved aerated concrete (AAC) masonry walls, another common building material, was carried out through experimental and numerical tests [53]. The same effect of wall thickness and boundary conditions was found for AAC wall as for the clay brick walls. The thickness of the walls agrees with elasticity theory that stiffness is proportional to wall thickness cubed. When a peak pressure of 10 kPa interacted with a 120 mm thick wall it lead to collapse, whereas the 180 mm and 240 mm walls survived. When studying the boundary conditions, the effect was similar to that seen with the masonry walls: more connections produce a much stronger structure [53].

The AAC walls in this study were much more liable to failure at lower pressures than clay brick masonry. The AAC walls were half the thickness of the clay brick walls, which has a significant effect. Even so, AAC walls are still significantly weaker. In addition, a major damage mode at higher gas overpressures was the combined failure and fragmentation of the AAC bricks themselves. This fragmentation would cause impact injuries on occupants.

The use of pre-stressing in construction is a common practice. Parisi et al. [54] investigated pre-stressing in the response of unreinforced tuff-stone masonry (TSM) walls to blast loads. The study looked at the out-of-plane behaviour of load bearing TSM walls from VCEs. A triangular pulse load where the rise time was half the total duration was used. Only near collapse damage was considered in order to simulate a worst-case scenario. Three levels of vertical compression were applied to the walls, at 10, 25 and 50% of the uniaxial compressive strength of TSM.

An increase in the pre-compression ratio from 10 to 25% led to a 130% increase in blast resistance for the impulsive loading asymptote and a 97% increase for the

Table 4 Pressures at which generic building structures fail in a VCE

Structure	Typical failure pressure (kPa)
Glass windows	2–7
Room doors	2–3
Light partition walls	2–5
50 nm thick breeze block walls	4–5
Unrestrained brick walls	7–15

quasi-static asymptote. Increasing to 50% axial load resulted in a doubling of the blast resistance to impulsive loading and approximately a 39% increase to quasi-static loading.

5.6 Weaker Component and Glass Windows

While global collapse of buildings may be found near the point of explosion, the loading and resulting damage will decrease with distance. This will result in lower levels of structural damage. However, the fracture of weaker components can result in the generation of debris. The failure pressures for some of the weaker components commonly found in buildings are shown in Table 4.

Harris et al. [55, 56] conducted experiments to examine the hazard presented by glass fragments. Experiments were carried out in a concrete bunker, where the open end was closed with steel cladding into which different sized windows could be fitted. The blasts in the bunker were created using stoichiometric air-gas mixtures inside weather balloons held in place within the bunker. The pressures within the bunker were measured and this allowed the breaking pressures of plain, patterned, and Georgian-wired glass to be obtained as a function of area and thickness. The glass exhibited two main failure modes: circumferential or radial fracture was produced by slow rising pressures, i.e. deflagration, while circumferential fracture was only seen when fast rising pressures were used, i.e. detonation. In many practical cases the failure mode most encountered is radial fracture, probably due to local loading conditions. Evidence of a mixed damage pattern, of radial and circumferential, suggests high pressure blast damage from high explosives.

Panels that tend to fail flexurally do so as a function of the square of their dimensions, while there was a roughly linear relationship between the pressure causing failure and the glass thickness. Scaling laws that usually apply to materials such as brick walls are not followed in glass, which may be due to the more complex failure modes exhibited by many windows.

Fragments from broken windows are dangerous to personnel as the velocity of fragments from a 25 kPa blast can travel up 40 m s^{-1} [56]. Glass fragment injuries are one of the most common and far-reaching injuries following an explosion. Not only can these fragments cause human injury, but they can also damage surrounding buildings and property. Houses and flats in the UK use a variety of glasses from standard annealed float glass to toughened, laminated, and wired glass. The latter classes are much stronger than annealed float glass when it comes to impacts and blast loading [57]. The velocity and travel distance is heavily influenced by fragment area and shape, i.e. the air resistance.

Laminated glass is used on the outside of many modern buildings. It is composed of two or more panes sandwiching one or more polymer interlayers. This acts to increase the strength of the glass and reduce the likeliness of fracture [57]. This type of glass is used externally to protect the building from the elements and from external blast loads. Since this type of glass is tough to fracture it means that in a confined VCE in a building the blast pressures will not be able to vent as quickly but is less likely to injure personnel outside the building [58].

Kummer [59] developed a model to evaluate the risk posed by window fragments resulting from a blast. Kummer states that many lethality/injury models require many inputs which are not always available after a blast, i.e. window thickness. Many also do not consider the impulse of the blast and only report or estimate the peak pressure. The window types considered in Kummer's model are small ($<1 m^2$), medium ($1–3 m^2$) and large ($>3 m^2$). Thickness of 4–6 mm, dual pane, float glass (i.e. not toughened or laminated), and modern (<40 years old).

6 Mitigation

The ability to mitigate a blast effect is strongly dependent on local conditions. Buildings will already be in place and built according to a range of practices, often in a mixed and crowded environment with roads, walls, and neighbouring buildings. A fundamental consideration is the needs of the occupants. Radically different solutions are needed if the building is occupied by an interim government in an on-going conflict, compared to a civilian building which may at some point be in the path of a tornado.

It may well be the case that a location has only very limited options and the occupants cannot simply move away. Faced with a complex situation, it is useful to consider actions in broad principles.

(a) Assess the threat and the initial position:

  (1) What is the likely threat: deliberate, accidental or natural?
  (2) How and where is threat going to arrive, what is its likely size, can it be kept at a distance?
  (3) Is it possible to move out of the way of the threat?
  (4) Can the site be cleared to allow larger open spaces and fewer objects that can turn into fragments and projectiles?

(b) To assess blast effects, more detailed considerations can be undertaken and some calculations made of the overpressure and probable effect:

  (1) Is the threat more like a deflagration or a detonation?

(2) What is the TNT equivalence of the likely threat and what are the pressures generated?
(3) Can access to the building/site be restricted?
(4) Can roads be blocked and obstacles introduced to prevent large devices being driven onto the site or kept away from perimeter walls?
(5) Are there drains, underground structures where devices can be introduced?
(6) Overlooking buildings, countryside?

(c) Analyse the existing structures:

(1) What structures exist on the site and of what materials and thickness?
(2) Where are the likely weak points for progressive collapse?
(3) Are there detailed plans available from the maintenance department? (More likely depending on the building status/complexity)
(4) Can strengthening be done to critical elements?
(5) Is improved venting a possibility? (For internal blasts)
(6) Can the internal layout be altered and room use considered?
(7) Reduction of flammable materials and pressured gases on site?

Computer modelling can provide broad answers to many of the above questions. Materials such as spray-on foams or Kevlar-reinforced 'wallpaper' can be used on brick walls. Window glass can be removed and replaced by perforated metal sheets or strengthened weighted curtains used.

One feature of blast is that it can roll over hills and move around corners, something that projectiles cannot easily do. Therefore stand-off distance and the use of energy absorbing materials may be of greater effect than simple strong walls.

Many industrial accidents have been caused by the transfer of solvents or volatile materials. Static charges, leaking pipes or, in the case of the Buncefield explosion in 2005, the presence of a hedge, which may trap and prevent flammable gases from dispersing have all contributed. Explosive production plants incorporate weak walls and blow off roofs to prevent significant pressure build-up and allow a planned collapse. The presence of earth banks and limiting quantities of material and staff time in hazardous zones form a normal work environment.

Newer infrastructure can be blast–hardened from the design stage through the use of appropriate building codes and material specifications. Analysis of the proposed structural load-bearing sections can be adapted to prevent or reduce the effect of progressive collapse.

Medical research and trauma care are ensuring a much greater survival rate from weapons and terrorism, and are informing practice in the civilian sector. The use of dedicated trauma centres and concentration on the 'golden hour' immediately after injury have produced marked improvement in survival at both short- and long-term levels. This discussion is now moving from one of survival to one of quality of life.

7 Summary

This chapter has shown the complexity of blast research in terms of blast effects on a wide variety of materials, the change in their properties, and failure modes based on the accelerations produced by explosion. This chapter also indicates the difficulty of understanding the precise effects in terms of the materials and the three-dimensional world. However, by dividing complex situation into smaller parts it becomes more tractable.

An overview of the basis of explosive technology and some of the basic processes relevant to the blast process have been presented. The detonation wave, stress transmission, fragmentation, the ejection of sand/soil, expansion of the product gases, and flash heating have been addressed.

The complexity of the mechanical processes and the resistance of the human body can result in injury patterns that show effects and pathologies that evolve over time from the immediate blast or impact to longer term chronic effects. It is becoming clearer that it is possible to adopt an interdisciplinary approach that can bridge the vital mechanical–biological gap in our knowledge.

Buildings and vehicles are complex structures, often targeted or subject to accident. Much research has indicated the limits of common construction materials under explosive loading. Newer materials should be studied with respect to their intended function but also to the effects of blast and impact if they are to be used in key functions in areas of extreme natural phenomena, placed in conflict areas or in key industrial sites.

Future research in this area will need closer interaction of modelling, experiment and theory and attention to the various failure modes. There is much to be done.

References

1. Ritenour AE, Blackbourne LH, Kelly JF, McLaughlin DF, Pearse LA, Holcomb JB et al (2010) Incidence of primary blast injury in US military overseas contingency. Ann Surg 251(6):1140–1144. https://doi.org/10.1097/SLA.0b013e3181e01270
2. Magnus D, Khan MA, Proud WG (2018) Epidemiology of civilian blast injuries inflicted by terrorist bombings from 1970–2016. Def Technol 14(5):469–476
3. O'Neill C, Robinson AM, Ingleton S (2012) Mitigating the effects of firebomb and blast attacks on metro systems. Procedia Soc Behav Sci 48:3518e27
4. CO-Gas safety. Co-gas safety's statistics of deaths and injuries. http://www.co-gassafety.co.uk/wp-content/uploads/2016/04/CO-GAS-SAFETY-STATISTICS-ON-DEATHS-AND-INJURIES.pdf.
5. Stennett C, Gaulter S, Propellants JA (2020) An estimate of the TNT-Equivalent Net Explosive Quantity (NEQ) of the Beirut port explosion using publicly. Available Tools and Data, Explosives, Pyrotechnics, October 2020. https://doi.org/10.1002/prep.202000227
6. Cooper PW (1996), Explosives Engineering, (Wiley-VCH)
7. Fordham S (1980) High explosives and propellants. Pergamon Press, New York
8. Bailey A, Murray S (1989) Explosives, propellants and pyrotechnics. Brassey's, UK
9. Robert M, Pachman J (2013) Primary explosives. Springer

10. Proud WG, Field JE, Milne A, Longbottom AW, Haskins PJ, Briggs RI, Cook MD (2003) The detonation of NM/Al mixtures. In: Proceedings of fifth international symposium on high dynamic pressures, Commissariat à l'Energie Atomique, Paris, vol 1, pp 135–141
11. Grady D (2006) Fragmentation of rings and shells: the legacy of N.F. Mott. Springer
12. Gurney RW (1943) The initial velocities of fragments from bombs, shells and grenades. BRL Report 405
13. Proud WG (2013) The physical basis of explosion and blast injury Process. J.R. Army Med Corps 159 (Supp I):i4–i9. https://doi.org/10.1136/jramc-2013-000030
14. Bowen IG, Fletcher ER, Richmond DR, Estimate of Man's Tolerance to Direct Effects of Air Blast. Report: DASA-2113, October 1968.
15. Richmond DR, Yelverton JT, Fletcher ER, New Airblast Criteria for Man. Report: LA-UR-86, August 1986, Los Alamos National Laboratory, USA
16. Sadwin LD, Michael M. Swisdak (2013) Reflected and negative phase airblast energy measurements. New Trends Res Energet Mater, Czech Republic, 325–330
17. Kinney GF, Graham KJ (1985) Explosive shocks in Air. Springer-Verlag, Berlin
18. Needham, CE Waves B (2018) 2nd edn. Springer
19. Chen A (2013) Structural response to vapour cloud explosions. PhD thesis, Imperial College London. https://doi.org/10.25560/23922
20. Eiband M (1959) Human tolerance to rapidly applied accelerations: a summary of the literature. Memo 5-19-59E, NASA, Lewis Research Center, Cleveland, OH, June 1959
21. Drapela P, Lorenzo R, Lampert S (2008) How to quantify the effects of non-lethal kinetic weapons. In: 24th international symposium on ballistics, New Orleans, LA, 23–27 September 2008, pp 1284–1290
22. Field JE (1998) Private communication, Cavendish Laboratory, University of Cambridge. Images taken with a camera supported by the EPSRC
23. Williams MR, Dunne LJ, Pearce B, Lawton B (2001) Creams for protection against skin burns in explosions. In: 19th international symposium on ballistics, Interlaken, Switzerland, 7–11 May 2001, pp 1035–1042
24. Pirlot M, Dyckmans G, Bastin I (2001) Soap and gelatine for simulating human body tissue: an experimental and numerical evaluation. In: 19th international symposium on ballistics, Interlaken, Switzerland, 7–11 May 2001, pp 1011–1017
25. Knudson PJT (1998) Comparison of 10% and 20% ordinance gelatin using 7.65 mm full metal jacketed bullets. In: 17th international symposium on ballistics, Midrand, South Africa, 23–27 March 1998
26. Kundsen PJT, Theilade P, Vigsnaes JS, Neilsen PT (1995) Applied wound ballistics: the case of the Danish 7.62 mm NATO bullet. In: 15th international symposium on ballistics, Jerusalem, Israel, 21–24 May 1995
27. Gray W, Weiss CE, Walker JD, Sponsel WE (2008) Computational and experimental study of paintball impact ocular trauma. In: 24th international symposium on ballistics, New Orleans, LA, 23–27 September 2008, pp 1260–1267
28. Newell N, Neal W, Pandelani T, Reinecke D, Proud WG, Masouros SD (2016) The dynamic behaviour of the floor of a surrogate vehicle under explosive blast loading. J Mater Sci Res 5(2);59–67
29. Ngo T, Mendis P, Gupta A et al (2007) Blast loading and blast effects on structures. EJSE 2007:76–91
30. Cormie D, Mays G, Smith P (2009) Blast effects on buildings, 2nd edn. Thomas Telford, London
31. Koccaz Z, Sutcu F, Torunbalci N (2008) Architectural and structural design for blast resistant buildings. In The 14th world conference on earthquake engineering, October 2008 pp. 12–17
32. Cullis IG (2001) Blast waves and how they interact with structures. J Royal Army Med Corps 147(1):16–26
33. Shi Y, Hao H, Li Z-X (2008) Numerical derivation of pressure–impulse diagrams for prediction of RC column damage to blast loads. Int J Impact Eng 35(11):1213–1227

34. Jarrett DE (1968) Derivation of the British explosives safety distances. Ann New York Acad Sci 152(1):18–35
35. Naito CJ, Wheaton KP (2006) Blast assessment of load-bearing reinforced concrete shear walls. Pract Period Struct Des Construct 11(2):112–121
36. toursbymawiyah. https://toursbymawiyah.wordpress.com/global-prayer-vigilcandlight-memorial-for-the-people-of-ayiti-peace-pole/. 21 December 2018 cited from Google
37. Siddam A (2014) Cascade failure analysis of electrical transmission lines using ADINA (Doctoral dissertation, Memorial University of Newfoundland)
38. http://www.wsdot.wa.gov/TNBhistory/3bridges/1940/Panel1Pic10.htm. 21 December 2018 cited from Google
39. https://kaiserpermanentehistory.org/tag/t-2-tanker/. 21 December 2018 cited from Google
40. https://www.fhwa.dot.gov/publications/publicroads/96fall/p96au17.cfm. 21 December 2018 cited from Google
41. Adam, Parisi, Sagaseta, Lu (2018) Research and practice on progressive collapse and robustness of building structures in the 21st century. Eng Struct 173:122–149
42. Lin TY, Burns NH (1963) Design of pre-stressed concrete structures. Technical report, Wiley New York
43. Pearson C, Delatte N (2005) Ronan point apartment tower collapse and its effect on building codes. J Perform Constr Facil 19(2):172–177
44. Levy M, Salvadori M (2002) Why buildings fall down: How structures fail. WW Norton & Company
45. Griths H, Pugsley A, Saunders OA (1968) Report of the inquiry into the collapse of flats at Ronan Point, Canning Town: presented to the Minister of Housing and Local Government. HMSO
46. Bob C (2004) Evaluation and rehabilitation of a building affected by a gas explosion. Progr Struct Eng Mater 6(3):137–146
47. Li J, Hao H (2014) Numerical study of concrete spall damage to blast loads. Int J Imp Eng 68:41–55
48. Wang W, Zhang D, Lu F, Wang S-C, Tang F (2012) Experimental study on scaling the explosion resistance of a one- way square reinforced concrete slab under a close-in blast loading. Int J Imp Eng 49:158–164
49. Smith P, Cormie D (2009) 3 Blast loading. ICE Publishing, pp 30–58
50. Dragos J, Wu C (2014) Single-degree-of-freedom approach to incorporate axial load effects on pressure impulse curves for steel columns. J Eng Mech 141(1):04014098
51. Wei X, Stewart MG (2010) Model validation and parametric study on the blast response of unreinforced brick masonry walls. Int J Imp Eng 37(11):1150–1159
52. Li Z, Chen L, Fang Q, Hao H, Zhang Y, Xiang H, Chen W, Yang S, Bao Q (2017) Experimental and numerical study of unreinforced clay brick masonry walls subjected to vented gas explosions. Int J Imp Eng 104:107–126
53. Li Z, Chen L, Fang Q, Hao H, Zhang Y, Chen W, Xiang H, Bao Q (2017) Study of autoclaved aerated concrete masonry walls under vented gas explosions. Eng Struct 141:444–460
54. Parisi F, Balestrieri C, Asprone D (2016) Blast resistance of tuffstone masonry walls. Eng Struct 113:233–244
55. Harris RJ (1983) The investigation and control of gas explosions in buildings and heating plant. E & FN Spon in association with British Gas Corporation
56. Harris RJ, Marshall MR, Moppett DJ (1977) The response of glass windows to explosion pressures. British Gas Corporation. I. Chem. E. Symposium Serious no. 49
57. Hooper PA, Sukhram RAM, Blackman BRK, Dear JP (2012) On the blast resistance of laminated glass. Int J Solids Struct 49(6):899–918
58. Larcher M, Solomos G, Casadei F, Gebbeken N (2012) Experimental and numerical investigations of laminated glass subjected to blast loading. Int J Impact Eng 39(1):42–50
59. Kummer PO (2004) Glass breakage and injury-yet another new model? Technical report. Bienz Kummer and Partner Ltd. Zollikerberg (Switzerland)

Risk Assessment, Resilience and Adaptability for Future CBRNE Security

M. Endregard and K. O. Nystuen

Abstract It is a collective public interest to seek to minimise the risks of adverse effects caused by CBRNE events. Hence, societies, the international community, groups and individuals aim to regulate, reduce and control CBRNE risks to achieve security and safety. The main challenges one faces when attempting to govern CBRNE risks in contemporary societies are complexity and uncertainty. Rapid development of the society in concert with huge and fast changes in technological and global trends, call for considerable changes to how we think about and conceptualise risks, as well as new approaches for governing risks. To be able to handle complexity, surprises and uncertainty it is necessary to develop a new holistic common mind-set for risk, safety and security, where risk governance is handled in close concert with cautionary, precautionary, robustness and resilience strategies.The chapter presents such strategies that may complement risk assessments and inform decision makers on governing risks to avoid or mitigate consequences of CBRNE crises.

1 Introduction

Chemical (C), biological (B), radiological (R), nuclear (N), explosive (E) incidents may harm the lives and health of people, animals, the environment or infrastructures. The cause may be exposure to toxic, infectious or radioactive compounds, explosions, or fission of nuclear materials resulting in dispersed radioactive fission products or direct irradiation, or in the case of nuclear weapons, an extreme shock wave and heat. As illustrated by concrete examples in chapter "The Changing Landscape", CBRNE crises may be intentional, i.e. terrorist attacks or sabotage, accidental or a result of natural disasters such as pandemics, hurricanes, flooding or earthquakes. CBRNE threats are an interwoven part of our modern societies, since we rely on

M. Endregard (✉) · K. O. Nystuen
Norwegian Defence Research Establishment (Forsvarets Forskningsinstitutt (FFI)), P.O. Box 25, NO-2027 Kjeller, Norway
e-mail: Monica.Endregard@ffi.no

K. O. Nystuen
e-mail: Kjell-Olav.Nystuen@ffi.no

P.D.E. Biggins and D. Chana (eds.), *CBRNE: Challenges in the 21st Century*, Advanced Sciences and Technologies for Security Applications,
https://doi.org/10.1007/978-3-031-17374-5_6

the use of chemicals, explosives and radioactive sources in industry, commerce, the transport sector and agriculture. Such compounds are subject to the dilemma of dual use. They are highly beneficial when utilised for the good of society but may cause harm if exploited to produce weapon materials. Export control regimes are in place to counteract dual use, for instance the Australia Group for chemical and biological compounds, the Nuclear Suppliers Group and the United Nations Security Council Resolution 1540. To eradicate all infectious diseases is impossible, although modern medicine and public health rid us of smallpox. New diseases will emerge, as illustrated by the recent COVID-19 pandemic. Nuclear power is an important source of energy in many countries. In addition, chemical and biological weapons arsenals still exist despite their ban through the Chemical Weapons Convention of 1993 and the Biological and Toxin Weapons Convention of 1972. Nuclear weapons are held by the five recognised nuclear-weapon states according to the Non-Proliferation Treaty, as well as by India, Pakistan and North Korea, which have all tested nuclear weapons. Israel's possible nuclear arsenal is subject to ambiguity since Israel neither confirms nor denies possession of such weapons. In conclusion, CBRNE threats to society persist.

It is a collective public interest to seek to minimise the risks of adverse effects caused by CBRNE events. Hence, societies, the international community, groups and individuals aim to regulate, reduce and control CBRNE risks to achieve a state of security and safety. This encompasses crisis prevention, emergency preparedness as well as developing capabilities for crisis management, mitigation of consequences and strategies for normalisation and learning from incidents. The institutional structures and policy processes that guide and restrain such activities constitute what is denoted *risk governance* in which a multitude of actors and processes lead to decisions [1]. A widely accepted definition of *risk* is that "risk refers to an uncertainty about and severity of the events and consequences (or outcomes) of an activity with respect to something that humans value" [2]. Values are for instance human lives and health, the environment and material assets.

The main challenges one faces when attempting to govern CBRNE risks in contemporary societies are complexity and uncertainty. Complexity entails a difficulty of predicting system behaviour based on the system's constituent parts. This is truly the case for society as a system and for many sub-systems such as cities, transport systems or nuclear power plants. Uncertainty applies to both the likelihood and possible causes of CBRNE events, magnitude of consequences, subsequent chain of events, as well as possible cascading effects caused by societal and systemic complexity.

Rapid development of the society in concert with huge and fast changes in technological and global trends, call for considerable changes to how we think about and conceptualise risks, as well as new approaches for governing risks. Chapter 1 introduces and provides an overview of different terms and views related to risk and security in the context of CBRNE threats in an uncertain world. However, to meet threats, risk and uncertainty, we need to be able to identify and handle the different aspects of risk in a structured and agile way. Otherwise, such an important matter remains just a subject of talk.

Although scholars have discussed and acknowledged an increasing complexity in society as a whole, traditional risk analysis approaches are still widely used without sufficiently taking complexity and uncertainty into account. Too often, analysts are not taking into consideration the rapid increasing complexity of modern societal functions and infrastructures, as well as the increasing complexity of the technology surrounding us. Many still restrict their thinking of risk to the probability of something happening and the associated consequences of that event in one or another form. At the same time, we are developing advanced security concepts based on resilience, apparently without any clear relation to the development of risk approaches. Clearly, we are struggling to reach a unified view about what risk and security really is about, and how we can operationalise risk and security concepts to achieve sufficient holistic security for society and enterprises.

To be able to handle complexity, surprises and uncertainty in governing CBRNE risks it is in our opinion necessary to develop a new holistic common mind-set for risk, safety and security, where the development of risk governance is handled in close concert with strategies of caution and resilience. In this chapter, we discuss challenges and present ideas on how to achieve this. The structure of the chapter is as follows:

- Risk concepts relevant to this chapter and their definitions
- Trends, developments, and complexity of todays and tomorrow's society, with implications as a result of technological and global trends and the development of new services and infrastructures
- Concepts of complexity and surprises
- The Fukushima disaster from a resilience engineering point of view
- How to better learn from accidents using a system-theoretic approach
- The cautionary principle in relation to risk management
- Concluding remarks

2 Risk Related Concepts and Definitions

The Society for Risk Analysis (SRA) has found that it is not realistic to arrive upon a unified set of definitions related to concepts within risk research. However, SRA has agreed on a glossary of definitions acknowledged by a committee comprising leading scientists within the risk research field [3]. The glossary includes several definitions for some concepts, thereby representing different perspectives. SRA also distinguishes between qualitative definitions of concepts and their metrics. In this chapter, we use terminology and definitions in accordance with guidance from the SRA. For our purposes, the following key terms and definitions are important for further discussions in the present chapter [3]:

> "*Risk* refers to an uncertainty about and severity of the events and consequences (or outcomes) of an activity with respect to something that humans value." Risk is expressed as (C, U), i.e. the consequence (C) of the activity with associated uncertainties (U).

> *Uncertainty*: For a person or group of persons, not knowing the true value of a quantity or the future consequences of an activity.
>
> "*Probability*: A measure for representing or expressing uncertainty, variation of beliefs, following the rules of probability calculus." There are three types of probabilities:

- Classical probability P(A) which applies in situations with a finite number of outcomes, equally likely to occur.
- Frequentist probability $P_f(A)$ which is the limiting fraction of times the event A occurs if the situation considered were repeated (hypothetically) an infinite number of times.
- Subjective (judgmental, knowledge-based) probability.

> A system is *complex* if it is not possible to establish an accurate prediction model of the system based on knowing the specific functions and states of its individual components.
>
> *Resilience* is the ability of the system to sustain or restore its basic functionality following a risk source or an event (even unknown)." SRA also includes the following resilience definition: "*Resilience* is the ability of a system to reduce the initial adverse effects (absorptive capability) of a disruptive event (stressor) and the time/speed and costs at which it is able to return to an appropriate functionality/equilibrium (adaptive and restorative capability).

The International Organization for Standardization (ISO) issued revised guidelines for risk management in 2018 [4]. ISO defines "risk management" as "coordinated activities to direct and control an organization with regard to risk", thus part of governance and leadership. The standard emphasises that "The risk management process should be an integral part of management and decision-making and integrated into the structure, operations and processes of the organization." This ISO standard outlines principles, a framework and a process for risk management, and points out that this should be an iterative process. Figure 1 shows the risk management process as illustrated by ISO.

This chapter focuses on the risk assessment part of risk management and other strategies that may complement risk assessments to inform decisions on risk treatment and acceptance. Risk assessment entails identifying, characterising and evaluating risk as input to support decision-making processes, while risk management and governance comprises balancing inputs from risk assessments with information from other processes such as cost–benefit analyses, precautionary concerns and societal values such as privacy and human rights. "Risk assessments support decision-making; it does not prescribe what is the best decision" [5].

3 Trends and Developments

The building blocks and abilities to understand, prevent, prepare for, handle and recover from CBRNE risks and events lie in the resources, institutions, infrastructures and vital societal functions of society. Hence, in principle, to govern CBRNE risks one needs a comprehensive understanding of society. However, the consequence of

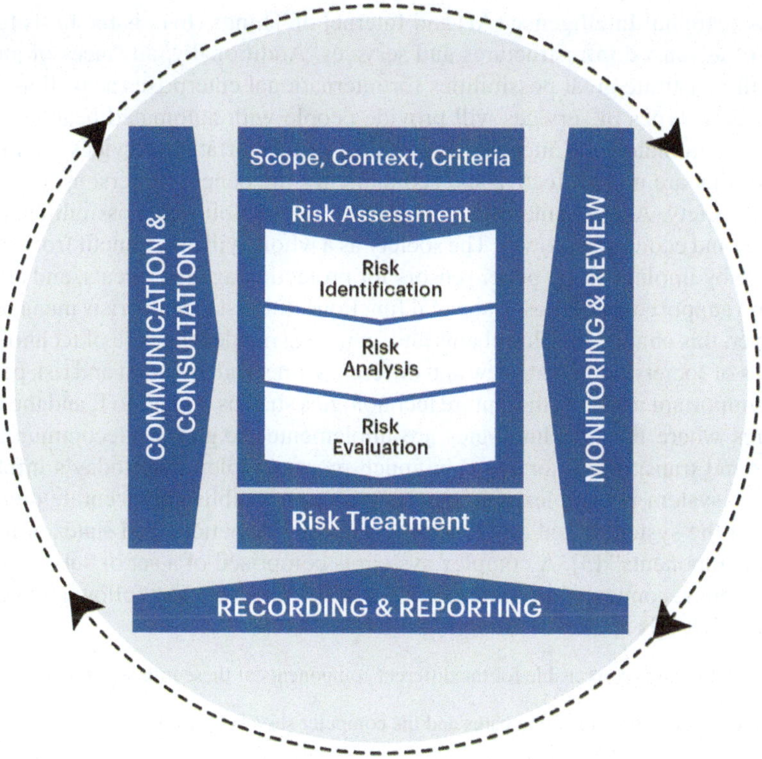

Fig. 1 The risk management process (Fig. 4 - Process from NS-ISO 31000:2018 is reproduced by Norwegian Defence Research Establishment (FFI) under licence from Standard Online AS August 2021. © All rights are reserved. Standard Online makes no guarantees or warranties as to the correctness of the reproduction. See www.standard.no.»)

current trends and developments is that nobody can fully understand how our society changes and the implications on safety, security and individuals' wellbeing. Our surroundings and the way we live are changing at a rapid pace to an increasingly complex world. It is ever more challenging to understand how the full spectrum of emerging trends, that also imply discontinuity in many areas, will affect us. The development in areas such as international security politics, social and economic conditions, climatic changes, and last but not least the technological development, have individually and together a huge impact on all risk factors in modern society. How we imagine threats are changing. How we perceive vulnerabilities are different. Last, but not least, how we perceive the values that we need to protect, spanning from the individual to the societal perspective, is rapidly evolving.

Most people acknowledge that the developments in information and communication technology will have a huge impact on society. Communication networks have an increasingly global footprint. Within a few years, there will be access to internet in most of the world. At the same time, the development of technological areas such as Artificial Intelligence (AI) and Internet of Things (IoT) is the basis for new forms of advanced infrastructures and services. Additionally, advances of automation will constitute great possibilities for international enterprises as well as for the public. New types of services will provide people with automated health and care services, automated and more energy-efficient transportation services, and energy services that are more effective and economic for the benefit of persons, families as well as society. At the same time, enterprises will exploit new possibilities to gain progress and economic growth. The society as a whole will also benefit from this, for example by implementing better sensors for protection against threats, and effective decision support capabilities in societal functions, for instance in crisis management. However, this change in value chains and an increasing dependence of technology in all areas of society will create new and different vulnerability, threat and risk patterns.

An important aspect is that future technologies such as AI and IoT, and the infrastructures where these technologies are implemented, e.g. 5G telecommunication, energy and transport sectors, will be much more complex than today's infrastructures. "A system is complex if it is not possible to establish an accurate prediction model of the system based on knowing the specific functions and states of its individual components" [3]. A complex system is comprised of a set of sub-systems or sub-activities, connected in some known or unknown way. The following questions illustrate that this applies to emerging infrastructures:

Who owns and is responsible for the different components of these infrastructures?

Where are the communication paths and the computer servers located?

How are different types of technology being instantiated into systems and services?

Who is programming the functions of the services, and who is using the service, and for what purposes?

All are important questions to ask, but increasingly more difficult to answer. This is about using and tightly integrating services from infrastructures like electrical power supply grids and telecommunications into critical societal functions like health care, food delivery chains and so on. Therefore, this concerns our most critical value chains, and a key question is how can we manage risk in this context?

At the same time, when performing risk analyses, a traditional methodological apparatus is still often employed. In general, the basis for these approaches is the understanding of risk as a combination of probability of an unwanted event and the associated consequences. The risk management process is often hierarchic and based on the idea that a comprehensive system overview is achievable, and that the system can be broken down into sub-components. Further, the assumption is that there is a linear connection between the function of each of the sub-components and the function of the entire system, and that the system is static. Increasing complexity and interconnected value chains are manifestations that the underlying prerequisites for traditional risk management approaches are invalid.

Today, we have roughly two ways to handle complexity in a risk context. The most obvious way is to try to meet complexity with knowledge. However, this has proved to be extremely challenging, just because sufficient knowledge is increasingly difficult to obtain. The other way is to narrow and simplify the risk analysis to factors that are well known. Such a delimitation also requires substantial knowledge, especially regarding the value chains that are involved. With the development of complexity, both strategies for risk analysis show to be challenging.

Future risk management and governance will be challenging since necessary understanding will be hard to obtain. This relates to all aspects of the value chains involved, possible vulnerabilities, as well as potentially relevant threats against vulnerabilities and assets. Most of all, there is a danger of sub-optimalisation instead of a more holistic approach to security, where the traditional approaches to risk in the worst case may end up being counterproductive to risk management and governance. It is therefore necessary to explore new ways and mind-sets about risk. Recent research calls for combining and balancing several strategies. Jensen and Aven [6] propose to combine risk information with cautionary, robust and resilient measures, and not rely solely on risk assessment for decisions on how to prevent and handle risks, thus combining various methods and approaches within risk research [6].

CBRNE events, both intentional and unintentional, are fortunately quite rare, but the consequences may be grave. Endregard et al. (2011) give some examples [7]. The most severe nuclear disasters to date include the reactor meltdown on Three Mile Island in 1979, the Chernobyl nuclear reactor explosion in Ukraine in 1986 and the tsunami flooding of the Fukushima Daiichi nuclear reactors in Japan in 2011. Chemical industrial disasters include the Seveso chemical plant accident in Italy in 1976 and the Bhopal gas leak disaster in India in 1984. Pandemics are recurring events. In 2003 the outbreak of the Severe Acute Respiratory Syndrome (SARS), a viral respiratory disease caused by a coronavirus, was first reported in Asia, then spread to several continents and numerous countries. In 2009, the H1N1 influenza pandemic originated from Mexico and subsequently spread to most countries. However, these two pandemics cannot compare with the devastating death tolls and other impacts of the COVID-19 pandemic, which is the most severe pandemic in a century.

What strategies are needed for society and organisations to assess risks in general, and CBRNE risks in particular, informing decision makers to manage and govern risks to avoid or mitigate consequences of natural disasters, epidemics, accidents and terrorism? In the following, we will present challenges and propose some strategies to address them, based on published work and some recent advances in the risk science domain. Several of these scholars have referred to the Fukushima disaster to exemplify various aspects, challenges and approaches to identify, characterise and assess risks. Operation of nuclear power plants and the Fukushima disaster are used as examples throughout this chapter.

4 Complexity

Fundamental issues related to risk assessment approaches is amongst other papers based on the early work of Kaplan and Garrick from [8]. A risk analysis seeks to answer three questions:

- What can happen, i.e., something going wrong?
- How likely is it that it could happen?
- If it happens, what are the consequences?

Based on these questions, Kaplan and Garrick introduced the so-called risk triple in order quantitatively to express risk. They introduced risk as a set of triplets of scenarios (s), associated probabilities (p) and consequences (c). They highlight that risk involves both uncertainty and some kind of loss or damage, where probability is an estimate of uncertainty. An important aspect is that they *reject the idea that risk is probability times consequence*, as emphasized by Aven [5].

Complexity in socio-technical systems, societal functions and society as a whole, challenges traditional risk assessment approaches based on the assumption that risk associated with the systems sub-parts gives the overall risk picture. Jensen and Aven conclude that considering risk expressed by consequences and probabilities provides too narrow a perspective for complex systems [6]. They have studied complexity in a risk context and state: "Complexity is linked to the difficulty of predicting system behaviour based on the system's constituent parts." and "There are risks related to these systems, as the consequences of the activities generated by the systems are uncertain" [6].

Jensen and Aven use the definition of complexity by Johansen and Rausund as point of departure [9]:

> Complexity is an acknowledgment of limitations in the understanding of the following:
>
> 1. The sociotechnical system in all its operational contexts;
> 2. How risk can be assessed, based on the available knowledge and the assumptions about the system elements. An element here is a wide term that refers to individual inputs, components, processes, hazards, events, conditions, and so on.

Aven and Jensen have proposed a risk framework suitable for complex systems based on the SRA's definition of risk, i.e. "risk is uncertainty about and severity of the consequences of an activity with respect to something that humans value, i.e. health, environment and assets." If we use a nuclear power plant as example, the activity is the operation of the nuclear power plant. The values are human lives and health, the environment and material assets. The consequences are results of future operations on these values, which are uncertain, thus risk is (C, U) as previously mentioned.

Identifying hazards and threats towards the overall system function or values such as life and health is usually based on system-level data, observations and statistical data, but relevant knowledge may not be available. For rare events data and observations are scarce. In a design phase of a system, system performance data are non-existent. In addition, a main problem in complex systems is the lack of linearity, meaning that causal links are more complicated than single chains.

Uncertainty analysis is a central part of assessing and describing risk. Probability is the most common uncertainty measure. Jensen and Aven reject the applicability of frequentist probability and probability models for complex systems because frequentist probabilities require some stability. Complex systems are, however, by nature dynamic, i.e. change over time, adapt to certain situations, emerge and evolve. Subjective (knowledge-based, judgmental) probabilities can always be assigned. Such probabilities describe the assessor's uncertainty or degree of belief, conditional on the assessor's knowledge. The strength of knowledge is thus important to describe. The following criteria can be used: the amount of relevant system data/information, justification of assumptions made, agreement among experts, understanding of the phenomena involved, and the degree to which the knowledge has been thoroughly examined. Jensen and Aven arrive upon a new definition of a complex activity:

> The complexity of an activity is an acknowledgement of limitations in how the risk for the activity can be precisely described, based on knowledge about the individual sub-activities of the system. [6]

Common properties of complex systems are large numbers of parts interacting in a non-simple way, nonlinearities, emergence and adaptability. "The knowledge of the overall activity (system) is poor, even if we have strong knowledge at the sub-activity (sub-system) level" [6]. Traditional risk assessment and management approaches are based on models of the sub-activities, but not the overall activity or system, resulting in strong limitations and hampering the related risk management. Therefore, Jensen and Aven conclude that new approaches are needed in addition to risk assessments. These could be resilience engineering as proposed by Hollnagel [10] or approaches based on system-theory approach [11]. Subsequent sections illustrate what these approaches offer in addition to traditional risk assessments.

One way of thinking is to increase system knowledge, and in this way make the system less complex or even non-complex, but this is not feasible in most cases. Jensen and Aven argue to acknowledge complexity and instead shift the focus toward the risk management. This means to use additional strategies to risk assessments of complex systems to enable decision-makers to balance various concerns. Jensen and Aven conclude that "for complex systems, we need to balance risk information with cautionary, robust and resilient measures. Risk assessment alone should not be used to prescribe what to do" [6]. Three common strategies are used to manage risks: risk informed, cautionary/precautionary strategy of resilience and robustness and the discursive strategy.

5 Surprises

Since the famous book of Taleb in 2007 and his second edition in 2010 entitled "The Black Swan", the concept of *black swan events* has become widely used to describe events that come as a total surprise. The metaphor stems from the discovery of black swans in Australia by the Europeans. Until then, they were convinced that all swans are white. Taleb assigns the following attributes to a black swan event: rarity, extreme impact and retrospective (not prospective) predictability [12]. Rarity refers to the event being an outlier, outside the area of expectations. Unlike the discovery of the bird, Taleb adds the characteristic of extreme impact. In addition, people have a tendency of finding explanations for the event after the fact, claiming that the event could have been predicted.

Aven highlights that whether an event is denoted a black swan or not is dependent on who makes the assessment and when. Aven defines a black swan as "[...] a surprising extreme event relative to one's knowledge/beliefs, [...]" [13]. He distinguishes between three types of black swans [13]:

- (a) Unknown unknowns, which are events that were completely unknown to the scientific environment
- (b) Unknown knowns that are not on the list of known events by those making the assessment, but known to others
- (c) Events that are on the list but judged to have a negligible probability of occurrence and thus are not believed to occur

Aven provides the following examples of the three types [13]: an example of the first type is the effects of the thalidomide drug that caused severe birth defects in children. An example of the second type is the September 11 attack. An example of the third category is the tsunami that destroyed the Fukushima Daiichi nuclear power plant. A tsunami of this magnitude was removed from the list of possible scenarios because it was deemed too unlikely.

An additional image used for a surprising event of high consequence is that of a *perfect storm*, which denotes an event caused by a very rare conjunction of events [14]. Its origin is a storm in the North Atlantic in 1990 killing 12 people. The storm was a conjunction of three types of storms, one starting over the United States, a cold front from the North and a tail of a tropical storm from the South. All three types of storms were known to occur regularly, but the combination was very rare. Another well-known illustration of an unlikely combination of events is the alignment of parallel slices of Swiss cheese by Reason [15]. If the holes align, a loss or failure will occur. This is known as the Swiss Cheese Model. Hence, surprises may stem from unlikely conjunction of naturally occurring events or concurrent multiple accidental failures in a system. Paté-Cornell (2012) points out that "perfect storms" mostly involve aleatory uncertainties, i.e. randomness of rare, but known events, while "black swans" represent epistemic uncertainty or lack of fundamental knowledge [14]. Most real scenarios involve both types of uncertainties.

According to Paté-Cornell, the Fukushima Daiichi nuclear power plant disaster in Japan in March 2011 is a result of a rare conjunction of known phenomena for which the probability of their conjunction was severely underestimated [14]. A 14 m high tsunami wave created by an earthquake of magnitude 9 (on the Richter scale) flooded the power plant causing the disaster. This type of event had not occurred in recent times but had happened twice in history. However, the nuclear power plant design did not account for this, only for a maximum wave height of 5.7 m. Recent earthquakes founded the basis for design decisions. We will return to this case in the next section.

A special class of surprising events are intentional acts of crime, such as terrorism and sabotage. Here additional complicating factors apply. Dillon et al. [16] emphasise five key challenges associated with terrorism risks [16]. The risk is dynamic. Terrorism comprises low-frequency high-consequence attacks by strategically thinking human beings who adapt and alter targets and *modus operandi.* It is a race between the attackers and defenders. The resources to eliminate all threats or comply with all standards are limited, thus the challenge is to prioritise and allocate resources. There are a large number of possible attack scenarios and security options, thus difficult to assess the overall effect of mitigating measures. There is little available data of past events, leading to uncertainties in estimates of the risk. Finally, risk-based decision-making is a subjective endeavour to answer questions related to risk acceptance and what measures to reduce the risks are sufficient.

It has been claimed that predictability is the main difference between unintentional events (accidents and natural disasters) and intentional incidents like terrorism and other crime. However, lack of statistical data for all events that are highly unlikely, leads to large uncertainties. As explained earlier, probability is the most common way to express uncertainty and variation of beliefs. The SRA distinguishes between the frequentist probability and subjective probabilities. Frequentist probability is the fraction of times the event would occur if the experiment could be repeated infinitely under similar conditions. Subjective (judgmental, knowledge-based) probabilities are assessed based on degree of belief and available knowledge and experience of the assessor. "For unique events, like the occurrence of a serious terrorist attack in the next 10 years in a country, or the occurrence of a major nuclear accident in the next 30 years, there is no alternative to the use of subjective probabilities. Repeating the situations to produce frequentist probability does not make sense" [5]. For all types of surprises, a frequentist probabilistic approach is thus not advisable. Quantitative risk assessments based on frequentist probabilities can according to Maal et al. (2017) introduce a false sense of safety, thus be directly harmful [17]. In the case of rare events, whether they are safety or security related, quantitative methods should be used with care.

How to account for extreme and highly unlikely events in risk management is a general challenge, and of high relevance for CBRNE incidents due to their rarity though possible high consequences. Such events represent major challenges in risk management. There is a danger that such events are excluded entirely from consideration because they are deemed so unlikely.

According to Aven, there are three strategies to manage risk [13]:

(i) The risk-informed strategy includes treatment of risk, avoidance, reduction, transfer and retention based on the use of risk assess ents;
(ii) The cautionary/precautionary strategy, a strategy of robustness and resilience, which highlights features such as containment, constant monitoring, research to increase knowledge and development of substitutes;
(iii) The discursive strategy involves measures to build confidence and trustworthiness through the reduction of uncertainties, clarification of facts, and involvement of affected people, deliberation and accountability.

He argues that when uncertainties prevail, risk-informed strategies should be combined with cautionary, robust and resilient approaches. Both Hollnagel's approach of resilience engineering (see Sect. 6) and Leveson's STAMP are examples that are examined in the next two sections, followed by Aven's cautionary principle.

6 The Fukushima Disaster: Could Resilience Engineering Have Helped?

The disaster at the Fukushima-Daiichi nuclear power plant in Japan is an illustrating example where a traditional and one-sided approach to risk management failed. Hollnagel and Fuijita [18] have assessed the Fukushima disaster from the view of resilience engineering, which "… offers novel ways to confront the puzzles of complexity, interconnectedness, system of systems, and ultra-high reliability" [18]. We use their analysis and results to illustrate some key aspects of resilience engineering. The authors advocate that resilience engineering as a complementary approach to risk assessment could perhaps have helped withstand the consequences of the tsunami thereby avoiding the loss of nuclear power plants.

On 11 March 2011, a gigantic magnitude earthquake (measuring 9.0 on the Richter scale) occurred 130 km east of Oshika Peninsula of Tohoku. It caused a tsunamic wave extending 500 km with a width of 200 km, rising to the height of 40 m above sea level when hitting land. The Fukushima disaster was devastating. About 20 000 people either lost their lives or are still missing. The infrastructure damage to buildings, roads, railways, electricity, gas and water supplies was catastrophic. After the nuclear power plant lost external electricity supply due to the earthquake, the emergency diesel generators started. However, the diesel generators were located underground, thus were flooded with seawater when the 14–15 m high tsunami wave hit the plant. The total loss of electrical power caused three reactor cores to melt, and the release of massive amounts of radioactive materials. A few days later three reactors exploded due to hydrogen build-up inside the reactor vessels resulting in additional enormous releases of radioactive materials.

Hollnagel has introduced the following definition of resilience:

> Resilience is understood as the intrinsic ability of a system to adjust its functioning prior to, during, or following changes and disturbances, so that it can sustain required operations under both expected and unexpected conditions. [19]

He states that four key interrelated capabilities are necessary for a system or an organization to be resilient: to respond, to monitor, to learn, and to anticipate.

- Responding entails the ability to respond to regular and irregular incidents, perturbations and opportunities by planned actions according to plans and procedures or by adjusting the course of action to the situation. Hollnagel and Fujita [18] emphasize that responding requires a state of readiness and availability of resources, i.e. people, tools and materials. The foundation for pre-defined responses is both learning from experiences and anticipation about the future.
- Monitoring is to ensure constant awareness of, and seek knowledge about, near term changes affecting the system thereby calling for actions, as well as monitoring the system performance itself.
- Learning means to analyse experiences and events, understand them, identify the lessons learned and learn from the right lessons.
- Anticipating means to be able to anticipate future conditions, hazards, threats, developments and possible disturbances.

According to Hollnagel and Fujita [18] the nuclear power plant system failed in performance due to a combination of inadequate engineering anticipation or risk assessment during the design in combination with inadequate response capabilities. There is a close dependency between the ability to anticipate and the ability to respond since preparing a response requires advance consideration.

The purpose of design-related anticipation is to consider what could happen which can seriously damage the plant's integrity and function. This encompasses three types of threats:

- Regular threats occur frequently, thus a standard response is in place.
- Irregular threats are rare and vary, consequently a standard response has not been established.
- Unexampled events are unimaginable, unexperienced extreme events.

The Fukushima disaster falls in the latter category. The probability of an earthquake was known to be very high, but the actual earthquake was much larger than the design basis. The tsunami was twice as large as assumed. The protective wall was only designed to withstand a 5.7 m wave, not 14 m. The tsunami considerations were based on historic cases, but an even larger tsunami was overlooked.

Risk assessment constitutes the formalised way of anticipation. Experts generally agreed that the plant design was perfect. A total loss of electricity, thus also total loss of cooling capability, was unthinkable. A scientist had warned of a larger tsunami and recommended modifications to the plant, but this was deemed too improbable, thus not followed through. This is an example that the plant leadership relied too much on the probabilistic risk assessments. However, such risk assessments do not sufficiently reflect or deal with uncertainties or reflect the dynamics of severe conditions.

Hollnagel and Fujita state that "..., domain experts seem to believe that probabilistic risk assessments give sufficient foundation for risk management, hence the ability to respond" [19]. They disagree because severe circumstances require resources and actions beyond what the plant design can provide. The electricity at the Fukushima-Daiichi nuclear power plant was totally lost. The personnel tried alternative ways to provide electricity, including electricity trucks, but these failed to reach the plant in time, or were too heavy to lift by helicopter. This was not thoroughly included in the risk assessments. The key lesson learnt is that:

> [...] we need to think not only about how the plant should work and respond to regular threats, but also how it possibly could have or acquire the properties that would make it resilient. [19]

The assumed technical perfection hindered such thinking:

> [...], overconfidence in the experts' anticipation of what might go wrong limited the ability to monitor and respond—and, to some extent also to learn—hence impeded the development of resilience. [19]

The Japanese Nuclear Safety Commission admitted that the judgement to exclude the consideration of a station blackout was a mistake, which a working group had pointed out in 1993.

Hollnagel and Fujita conclude that formal risk assessment's trust established methods and models more than they should. They offer an acceptable trade-off between thoroughness and efficiency. However, because severe accidents are very rare, analysts easily become overconfident in the analysis methods. The strategy of "resilience engineering" advocates a constant sense of unease, to be mindful of what one does, as well as to counteract the overconfidence that is the side effect of the relative safety of nuclear installations.

Features of resilience engineering are to identify capabilities that a complex socio-technical system must have to perform acceptably in everyday situations as well as during accidents. It is a complement to other engineering methods and provides a contrast to classical risk assessment methods. It intends to show how things can go right, and prerequisites for this to happen, rather than just focusing on how things can go wrong.

Safety is not the freedom from unacceptable risks, but the ability to succeed during expected and unexpected conditions alike. It is necessary to describe the system as a whole, and to understand how the various functions are coupled and depend on each other. Large-scale socio-technical systems, such as nuclear power plants, have become so complex that they seriously challenge established safety measures. Resilience engineering may be of help to see the problems more clearly.

7 Learning from Accidents

Major accidents recur although they seem to have similar systemic causes. Why do accidents with similar causes keep happening? Why do we not learn from these events? Leveson argues that analysis methods do not discover the underlying causes of events [20]. Also, learning from experience does not work as it is supposed to and learning is happening in the wrong places [20]. This is due to inaccurate assumptions regarding today's systems, although accident analysis and learning from events are based on these assumptions. Hence, accident analyses and prevention efforts are of limited usefulness. Leveson proposes instead to use sophisticated models of causality based on systems thinking and systems theory. In addition, she argues that proactive analyses need increased emphasis. One such model is the system-theoretic accident model and processes (STAMP), proposed by Leveson. A key prerequisite is that the model reflects "the reality of today's complex socio-technical systems and not oversimplify the causes of accidents" [20].

Leveson states that "every major accident has precursors that might have been used to prevent the major loss" [20] She refers to the Bhopal chemical plant catastrophe and the Three Mile Island nuclear power plant incident as examples of accidents that were preceded by nearly identical events, but without loss. These were not properly investigated or analysed; thus changes were not made. In-depth analyses are time-consuming and costly, but what about the costs of catastrophes?

> The problem with the chain-of-events model of accident causation is that it oversimplifies causality and the accident process and excludes many of the systemic factors in accidents and indirect or non-linear interactions among events. It also does not hold for accidents where the cause(s) lies in the interaction among system components, none of which may have failed. [20]

Accident causation involves three levels:

- 1st level: Basic proximate event chain
- 2nd level: The conditions that allowed the events to occur
- 3rd level, top-level: Systemic factors that contribute to the conditions and events

Almost no accident causation models include systemic factors. Most accident models and accident analysis techniques suffer from the limitation of considering only the events underlying an accident and not the entire accident process. Processes control a sequence of events and describe system and human behaviour as it changes and adapts over time.

Leveson emphasises that: "Accident causation is a complex process involving the entire socio-technical system including legislators, government, regulatory agencies, industry associations and insurance companies, company management, technical and engineering personnel, operators, etc." [20]. The entire process needs to be examined; "Otherwise, only symptoms will be identified and fixed, and accidents will continue to recur" [20].

8 The Cautionary Principle

The definition of the cautionary principle is that "If an activity could be serious and subject to uncertainties, then cautionary measures should be taken, or the activity should not be carried out" [21]. Aven provides the following example of its use: German authorities decided to phase out their nuclear power plants by the end of

2022. Germany decided this in the aftermath of the 2011 Fukushima nuclear disaster. Half the members of the German ethical commission concluded that nuclear energy is not acceptable, emphasising the potential catastrophic consequences, independent of the likelihood of large accidents as well as the economic benefits to society. The other half used cost–benefit arguments to underpin their stand referring to other electricity sources being feasible, but with less risk as well as giving weight to the cautionary principle.

The well-known precautionary principle is a special case of cautionary principle. The SRA defines the precautionary principle as: "If the consequences of an activity could be serious and subject to scientific uncertainties, the precautionary measures should be taken, or the activity should not be carried out." Hence, the difference between the cautionary and precautionary principles are uncertainties versus scientific uncertainties, respectively.

The cautionary principle in relation to risk is important when the consequences of an activity can be extreme. Then caution is needed, and uncertainty and likelihood judgements affect the degree of caution. Caution is needed when consequences are sensitive to how the activity is realised, thus related to risk compensation and behavioural adaption. The cautionary approach comprises to be cautious to avoid problems, failures and losses later in case something unforeseen should happen. The principle highlights the case of scientific uncertainties, which calls for using the precautionary principle.

Examples of robust and resilience-based measures are safety barriers, different layers of protection, "defence-in-depth", redundancy, diversification, the ALARP (as low as reasonably practicable) principle, and so on. Hence, the cautionary principle is commonly adopted. An important point by Aven is that cautionary measures must always be balanced against other concerns such as efficiency, effectiveness and costs. The historical and social context and recent events strongly influence this balance. Therefore, different countries arrive at different conclusions regarding nuclear power, as well as different approaches for security measures against terrorism threats.

According to Aven, the cautionary principle meets a need in risk management, being an expedient, prudent and advantageous principle. "Analyses (such as risk assessments and cost–benefit analyses) have strong limitations in guiding decision-making in the case of large uncertainties" [21]. The cautionary principle represents a correction to analytical approaches.

9 Concluding Remarks

Currently, the world is struggling to deal with the most severe pandemic in a century, the coronavirus outbreak named COVID-19 pandemic originating in Wuhan in China in late 2019. Prior to the event many researchers had predicted and warned of such a global crisis. Due to global interconnectedness and travel patterns rapid spread of infections globally occurred, resulting in large death tolls and considerable strain on medical services around the globe. Strictly speaking, the pandemic is neither a black swan nor a surprising event as there were indicators of a potential pandemic due to previous disease outbreaks such as Ebola, SARS and H1N1. Nevertheless, the COVID-19 outbreak has had a large impact globally and seems to have caught national authorities by surprise.

What strategies can society and organisations utilise to assess risks in general, and CBRNE risks in particular, informing decision makers to manage and govern risks to avoid or mitigate consequences of natural disasters, epidemics, accidents and terrorism? How can society and organisations take low-probability, high-consequence scenarios into account? In this chapter, we hope to have raised awareness on important factors to consider when informing decision-makers on such risks.

Complexity, surprises and uncertainty call for sound risk governance approaches incorporating risk assessment, resilience and adaptability for future CBRNE security. The mind-set for dealing with risk and uncertainty should include risk assessment approaches in concert with strategies of caution and focussing on resilient measures. Traditional risk assessments often comprise a linear approach, identifying hazards and threats, probabilities and consequences. Traditional risk management approaches based on risk assessment and the ISO standards do not sufficiently account for complexity, surprises and large uncertainties of future CBRNE risks. Hence, there is a need to include cautionary, precautionary, robustness and resilience measures in order to meet uncertainties and potential surprises. In addition, discursive strategies in the case of interpretative ambiguity and normative ambiguity are important to reflect different views related to relevant values.

Through this chapter, we have put emphasis on different aspects of the challenge of developing and making adequate risk assessments as a fundament for risk management in the modern society. With emerging compound threats from CBRNE as well as emerging threats against our use of information technology and the connecting world, we need to achieve a better understanding of how to counteract the whole span of possible threats against systems as well as the society. A core challenge is that this has developed to be everybody's responsibility. The society and how the society are tied together, is becoming so compound and complex that any security organisation is no longer able to make sufficient holistic and at the same time detailed risk assessments as they did before. We must all take part in assessing the risk in our operations, as well as being a part in finding out what to do about it. Security and risk have to become an integral part of how we are doing business in government and industry, as well as in our private lives. However, this demands a huge advancement in how we develop knowledge on risk management and security. It also requires the development of advanced and at the same time sufficiently simple to use methods and best practices on how to be able to do this. In conclusion, this is about how we are securing our ways of living in a society surrounded by an increasing well of complexity.

References

1. Renn O, Klinke A, Van Asselt M (2011) Coping with complexity, uncertainty and ambiguity in risk governance: a synthesis. Ambio 40:231–246
2. Aven T, Renn O (2010) Risk management and governance: Concepts, guidelines and applications. Springer, Heidelberg
3. Society for Risk Analysis Glossary (2018) https://www.sra.org/wp-content/uploads/2020/04/SRA-Glossary-FINAL.pdf. Accessed 17 Jan 2021
4. International Organization for Standardization (2018) Risk Management—Guidelines. ISO 31000:2018 (E), 2nd edn
5. Aven T (2020) Three influential risk foundation papers from the 80s and 90s: Are they still state-of-the-art? Reliab Eng Syst Safety 193:106680
6. Jensen A, Aven T (2018) A new definition of complexity in a risk analysis setting. Reliab Eng Syst Safety 171:169–173
7. Endregard M, Breivik H, Heireng HS et al (2011). D2.1 Scenario template, existing CBRN scenarios and historical incidents. PRACTICE WP2 deliverable. Norwegian Defence Research Establishment, Kjeller, Norway https://www.ffi.no/publikasjoner/arkiv/d2.1-scenario-template-existing-cbrn-scenarios-and-historical-incidents. Accessed 17 Jan 2021
8. Kaplan S, Garrick BJ (1981) On the quantitative definition of risk. Risk Anal 1(1):11–27
9. Johansen IL, Rausand M (2014) Defining complexity for risk assessment of sociotechnical systems: a conceptual framework. Proc Inst Mech Eng Part O.: J Risk Reliab 228(3):272–290
10. Hollnagel E (2012) FRAM: the functional resonance analysis method for modelling complex socio-technical systems. Ashgate, Farnham, UK
11. Leveson NG (2011) Engineering a safer world. Systems thinking applied to safety. MIT Press, Cambridge, USA
12. Taleb NN (2010) The Black Swan. The impact of the highly improbable, 2nd edn Penguin Books, London
13. Aven T (2015) Implications of black swans to the foundations and practice of risk assessment and management. Reliab Eng Syst Saf 134:83–91
14. Paté-Cornell E (2012) On "Black Swans" and "Perfect Storms": risk analysis and management when statistics are not enough. Risk Anal 32(11). https://doi.org/10.1111/j.1539-6924.2011.01787.x
15. Reason J (1997) Managing the risks of organizational accidents. Ashgate, Aldershot
16. Dillon RL, Liebe RM, Bestafka T (2009) Risk-based decision making for terrorism applications. Risk Anal 29(3):321–335
17. Maal M, Busmundrud O, Endregard M (2017) Methodology for security risk assessments—is there a best practice? In: Walls, Revie, Bedford (eds) Risk, reliability and safety: innovating theory and practice. Taylor & Francis Group, London
18. Hollnagel E, Fujita Y (2013) The Fukushima disaster—systemic failures as the lack of resilience. Nucl Eng Techn 45(1):13–20. https://doi.org/10.5516/NET.03.2011.078
19. Hollnagel E, Paries J, Woods DD, Wreathall J (2011) (eds). Resilience engineering in practice: a guidebook. Ashgate, Farnham, UK
20. Leveson NG (2011) Applying systems thinking to analyze and learn from events. Saf Sci 49:55–64
21. Aven T (2019) The cautionary principle in risk management: foundation and practical use. Reliab Eng Syst Saf 191:106585. https://doi.org/10.1016/j.ress.2019.106585

CBRN Terrorism

M. A. Wilson

Abstract Many papers on CBRN terrorism begin the same way. First, they quote leading politicians or security figures saying how likely it is that there will be a CBRN terrorist attack in the near future and highlighting the potentially devastating consequences. This is then contrasted with the very low numbers of terrorist attacks that have employed CBRN weapons in the past. This disparity between the fear of CBRN attacks and the infrequency of their occurrence has been central to many of the papers published in this area. The primary concern of this review is therefore to explore the variation in expert threat assessment and consider the factors that can be used to gauge the likelihood of terrorist actors turning to CBRN weapons.

1 Introduction

Many papers on CBRN terrorism begin the same way. First, they quote leading politicians or security figures saying how likely it is that there will be a CBRN terrorist attack in the near future and highlighting the potentially devastating consequences. This is then contrasted with the very low numbers of terrorist attacks that have employed CBRN weapons in the past. This disparity between the fear of CBRN attacks and the infrequency of their occurrence has been central to many of the papers published in this area. The primary concern of this review is therefore to explore the variation in expert threat assessment and consider the factors that can be used to gauge the likelihood of terrorist actors turning to CBRN weapons.

Terrorist attacks using CBRN can vary from highly targeted attacks against specific individuals to attempts at mass casualty incidents. Whilst terrorist actors may direct their action toward casualties of whatever number, acts of sabotage against an installation (an infrastructure attack) may have been planned by terrorists purely to demonstrate their power, to generate publicity or to create fear. In this regard, it should be noted that threats of violence are also acts of terrorism. Although

M. A. Wilson (✉)
Institute for Security Science and Technology, Imperial College, London SW7 2AZ, UK
e-mail: margaret.wilson@imperial.ac.uk

P. D. E. Biggins and D. Chana (ed.), *CBRNE: Challenges in the 21st Century*, Advanced Sciences and Technologies for Security Applications,
https://doi.org/10.1007/978-3-031-17374-5_7

there are instances of sabotage made to appear accidental, these are rare, since most hostile actors wish their actions to be known, whether they are motivated by political concerns, personal grievances or financial extortion.

The definition of terrorism will be considered in some detail, however, there are one or two other basic issues that need to be highlighted at the start of this review. The reader will have already noticed that in this chapter, the E of CBRNE has been lost in the move to terrorism. In relation to terrorism, analysts only consider CBRN; because E, of course, is part of normal operations in terrorist activity. Bombings have always accounted for a large proportion of the recorded cases of terrorism [e.g., 1]. The types of explosives used in terrorism and the ways that they are applied have extremely important implications, but this is outside the scope of this chapter. We also need to make a distinction between CBRN and Weapons of Mass Destruction (WMD) [2]. Ackerman [3] highlights the fact that while CBRN incidents are often referred to inter-changeably with WMD incidents in the social science literature, they should not be taken to be the same thing at all. Not only is there no consensus on how many people or how much damage constitutes "mass destruction", but Chemical, Biological and Radiological can, and indeed are, used to cause harm to small numbers of highly targeted people. Conversely, traditional terrorist methods, such as bomb attacks, have resulted in very high numbers of casualties in the past.

It should be noted that CBRN attacks also have the capacity to inflict huge financial damage on the targeted country, not just because of behaviour change, but also because of the very high cost of post-incident clean-up operations, which can go well beyond the cost of traditional terrorist attacks such as explosions [2].

2 The Terrorist Context

The starting point for this analysis must necessarily be the nature of terrorism, because without a shared definition, experts cannot assume that they are all talking about the same kinds of action, and researchers cannot engage with one another in cumulative research.

Following a period of considerable debate, and after the publication of more than a hundred proposed definitions [4], researchers have settled on a definition of terrorism that is acceptable to most academic and government players. The first distinction to make is that we define terrorism at the level of the act, not the actor [5]. This overcomes any subjectivity in deciding which actors should be labelled terrorist; a distinction which is highlighted by the variability in designations from different countries [6]. Thus, a person does not have to be a member of a recognised terrorist group in order to carry out an act of terrorism, and not every act of violence carried out by members of terrorist groups constitutes terrorism. In order to qualify as an act of terrorism, the action must include violence or the threat of violence, which has the intention to generate fear in a wider audience than the immediate victims of an attack. Importantly, the intention of the action must also be to harness this fear in the public in order to put pressure on another party, usually a government, to bring about some kind of social, religious or political change. These three core components are not disputed by most government organisations or academic researchers who are concerned with understanding terrorism. The only variation between the different definitions relates to specific additional caveats for the inclusion or exclusion of cases, for instance, that the actions must take place outside a context of warfare, that the perpetrators must be non-state actors (thereby excluding all state actors from acts of terrorism), and whether the immediate victims are civilians, as compared to military personnel. Thus, different definitions may include any or none of these extra conditions.

When we apply this definition to the use of CBRN weapons it brings with it some inconvenient exclusions. The first is that if we exclude state actors and warfare, then many of the known historic uses of CBRN are not within our remit. The second is that without the purposeful aims of terrorism, that is the motive of political or other change, then acts of violence such as the Aum Shinrikyo attack on the Tokyo subway are also excluded. This type of action, designed only to cause harm to others with no projected political outcome, are perhaps best described as "random sociopathic acts" [7].

However, there is a third group of actions which have caused problems for terrorism research as well. Assassinations are defined as the intentional, targeted killing of a prominent person. Some authors have pointed out that since the motive for assassinations is often achieved once the victim is dead, and there is no use of fear in the general public to create pressure for change, then *by definition* an assassination cannot ever be considered an act of terrorism [e.g., 8]. Nevertheless, terrorist groups do use assassination on a frequent basis, and other authors have argued that they may well create fear in the general population [9], as well as fulfilling the longer term aims of terrorism relating to publicity, propaganda and recruitment. In addition, there are some forms of terrorism, such as animal rights and anti-abortion, that target named individuals such as specific scientists or medical practitioners. The threat of such targeted murders only differs from assassination in the level of prominence of the victim, and these attacks are designed to create a great deal of fear in the targeted occupations. What we do know about assassinations is that they very clearly have a wider impact, both on individuals and subsequently on society via the changes people make in their behaviour; from use of the transport systems [e.g., 10] to public protests [e.g., 11], politics and voter behaviour [e.g., 12] to investments on the financial markets [e.g., 13, 14]. This analysis will therefore consider assassination as a specific subset of terrorist action when it is carried out by or in the name of a terrorist organisation.

3 Missing Motives and Misdirection

Given the definition of terrorism, in order to decide whether an act fulfils the necessary criteria, it is clear that we need to know the motive. In the immediate aftermath of a potentially terrorist attack, law enforcement spokespersons and responsible journalists are therefore careful not to attach a terrorist label until more information is available. Often, even when terrorist actors have been killed at the scene, they will have left evidence of their motives, for example, in diaries, letters or social media posts. This relates very clearly to what is often referred to as the "symbiotic relationship" between terrorism and the media; an area of Terrorism Studies (TS) with its own quite substantial literature. Put simply, terrorism sells papers (or garners clicks), and one cannot bring about change if the public are not made aware of the perceived injustice [15]. Since many authors have highlighted that acts of terrorism are a form of communication, it is therefore surprising that a large proportion of the acts of violence assumed to be associated with terrorism are never claimed by, or in the name of, a terrorist organisation at all.

The possible reasons why terrorist attacks go unclaimed or are falsely claimed have been explored by a number of researchers [e.g., 16]. In addition to those cases that remain unclaimed, some are claimed by those who were not responsible, and some are claimed on others' behalf. It seems most likely that incidents claimed by those not responsible are simple attempts to take the "credit" without any of the risks involved in carrying out an attack. However, when attacks remain completely unclaimed, there are a number of possible explanations. Most terrorism researchers agree that certain types of attack can backfire on the organisation responsible, creating a "backlash" that can adversely affect the terrorists' cause. For example, attacks aimed at those who are not considered to be legitimate targets, such as children, are less frequently claimed [17, 18], and can result in loss of support from even the terrorists' own constituency (their supporters).

There are, however, some other hypotheses as to why an attack would remain unclaimed. Rapaport [19] suggests that there may be state sponsors behind some acts of terrorism, who wish to remain anonymous so as not to incur any reprisals. This view is also supported by Hoffman [20], who suggests that it may be the terrorist actors who fear reprisals, either against them personally or against the countries that are protecting them. Finally, Hoffman [21] pointed out that claims may be related to the level of competition between different terrorist groups in a particular region. In conflict zones where there is competition for supporters, we might expect higher rates of attributed attacks, whereas in other areas, such as the IRA's activities on the UK mainland, there may be less necessity to formally claim an attack. One interesting area of research that has so far received little attention is the empirical prediction of claims from observable qualities of the attack, such as the nature of the target and the number of casualties [see for example 22].

Finally, there are some incidents which are claimed on behalf of a rival group, and these certainly appear to be an attempt to put the blame on to someone else. Here again, it is likely that these will be the kinds of attack that are perceived to have been unsuccessful in some way. In relation to these hypotheses, it seems likely that CBRN attacks would be among those most likely to be received badly. Not only may they be perceived to be indiscriminate in their targeting, but they have the potential to affect many civilian members of the terrorists' own constituency. This is certainly one of the explanations that have been proposed as to why CBRN weapons are used so infrequently by terrorist actors, and this will be considered in more depth later in the Chapter.

Thus, any study of CBRN terrorism is hampered by cases where we are missing the motive and therefore cannot assume that the incident was an act of terrorism. In addition, it may also not always be certain *which* terrorist organisation may be responsible for a particular act, since no-one, more than one, or the wrong set of perpetrators may claim it. State actors may disguise their involvement by hiding behind terrorist actors, and individual actors may never be caught or may be killed at the scene. This potential misattribution or misdirection has implications for empirical studies of which types of terrorist actor might be inclined to attempt CBRN attacks in the future.

There is one final form of misdirection that we need to consider, which is where deliberate attacks are disguised as accidents or naturally occurring incidents. These types of attempted "cover-ups" are relatively frequent amongst cases of product contamination and poisoning, for example, where a person has killed someone out of anger or revenge or has murdered a relative in order to claim insurance money [23]. Some of these cases also use improvised CBRN materials. It is possible that a terrorist group, fearing a backlash, may attempt to disguise their work as naturally occurring or accidental, but there are few existing terrorist cases that are suspected to have been disguised this way.

4 What is the Threat?

In order to assess the threat of CBRN attacks we have to take into consideration not only the motivation of terrorist actors but also their capability. While there are concerns about terrorist actors being supplied with weapons by a state sponsor [24], many analysts feel that most CBRN weapons are beyond the capability of most threat actors, either because they could not access the basic materials required, or because they would not have the technological knowledge to weaponize them [e.g., 25]. Nevertheless, it has been highlighted that, thanks to widespread publicity, there are some forms of improvised weapons that can be made from ordinary household materials, and numerous terrorist suspects have been found in possession of "cookbooks" downloaded from the internet.

4.1 Expert Opinion

One approach to understanding the potential threat from CBRN terrorism has been to conduct studies interviewing experts within a certain field to see whether there is any level of agreement as to the size or the nature of the threat. The results show some consensus and much variation. For example, Boddie et al. [26] conducted a study with 62 experts who all held positions relating to national security and life science. The researchers asked them to estimate the likelihood that a biological weapons attack resulting in harm to more than one hundred people would happen in any country in the next ten years. There was little consensus between the experts, however, the biological scientists in the sample consistently rated the likelihood as lower. The arguments given for why the likelihood was considered to be low were, the "technological complexities of developing a bioweapon, the difficulty in obtaining pathogens, and ethical and/or cultural barriers to using biological weapons" [26, p. 793]. For those members of the expert panel who did consider such an attack possible, the most likely biological agent to be used in weapons was considered to be biological toxins, e.g., ricin.

In relation to expert judgements of threat, Koblentz [27] identifies three perspectives. There are "the optimists" who believed that terrorist organisations lack the intention or capability to mount such an attack, "the pessimists" who believe the situation is likely to change in the near future, and "the pragmatists" who believe that there may be terrorist aspirations towards these kinds of attack, but that only small-scale attacks are feasible.

Koblentz [27] points out that scientific debate in the area is hampered by the fact that experts with original positions on either side of the divide rarely change their point of view; and yet in a world defined by changing threats, we need our experts to reconsider their judgements based on changing circumstance. Interestingly, he goes on to outline the heuristics and biases that may distort expert opinion, pointing out that beliefs of both likeliness and unlikeliness are amplified by well-known cognitive processes that lead to poor probability judgements in all areas of life [28]. For instance, the availability heuristic holds that events that are easily called to mind are considered to be more probable than those that are not. Thus, our perceptions of the likelihood of various forms of terrorist attack can be directly affected by news coverage, especially of sensational or highly lethal attacks. Many have noted that experts of all kinds are far from immune to these cognitive distortions [e.g., 29].

Overall, studies of experts show that they have very different views on the likelihood of CBRN terrorism taking place. Nevertheless, even if only dealing in hypotheticals, there is some agreement in the predictions experts make about which groups might use which forms [26]. While the threat is perceived to be greatest from disgruntled or mentally ill individuals, if we are only considering terrorist actors, then the far right and religious extremists have been proposed as the most likely perpetrators [26, 30]. The reality of Al-Qaida's often-cited interest in WMDs has been the subject of careful analysis [31]. However, while there is some agreement over religious based terrorists being the most likely to seek to use weapons of mass destruction, or CBRN, citing their aim to annihilate "any and all perceived enemies" [24, p. 46], empirical studies contradict this. For example, Asal, Ackerman and Rethemeyer [32] examine the factors associated with a terrorist organisation aspiring to acquire CBRN weapons and find that those that do, are more likely to be well resourced and to be based in globalised countries. Although several other authors have made claims that religious organisations are amongst these, and whilst religious terrorism does have higher fatalities on average, there is no evidence from this study to support the link between religious motivations and CBRN pursuits.

Many authors conclude that those terrorist organisations who have the resources, expertise and connections to acquire CBRN weapons do not have the motivation. This stands in contrast to unaffiliated persons with personal agendas, such as "going down in history", who are likely to lack the capacity to put their fantasies into practice.

There are some commonalities in the conclusions drawn by groups of experts, such as the fact that biological attacks are considered more likely than nuclear, and that of the possible *terrorist groups* who may use CBRN weapons, religious extremists are considered the most likely perpetrators. However, beyond a few key conclusions, experts' judgements on the likelihood and nature of such events are remarkably wide-ranging.

Aside from expert opinion, there are two other possible approaches to assessing whether terrorists will use CBRN weapons. The first approach is an actuarial one; examining what has happened in the past, or what has been *attempted* in the past, maybe as an indicator of what people might aspire to in the future. The second approach is to consider what is known about the motivations of terrorism and draw conclusions based on a theoretical perspective. The approaches to assessing CBRN terrorist threat and their drawbacks are summarised in Fig. 1.

5 Studying the Past

Post 9/11, authors who choose to study patterns in terrorist activity, whether based on the weapons they use or a myriad of other variables, have a number of possible data resources at their disposal. Although there were some key databases being widely used by academic and government researchers in the 1980s and 1990s (e.g., Rand, ITERATE), the last twenty years has seen a number of new databases developed to cover both general and specific forms of violence. In relation to terrorist attacks in general, the most widely used database is The Global Terrorism Database (GTD), however, there are other data sets which may have specific relevance to CBRN. Big Allied and Dangerous (BAAD) and Failed Foiled Successful and Completed (FFSC) contain data relating to terrorists' aspirations to use CBRN weapons and the Monterey WMD dataset records both terrorist and criminal acts to include plots, hoaxes and threats. It should be noted that all of these databases have been put together from open-source data, mainly news reports. As such, they all acknowledge their potential biases and limitations, chief among these being the effects of Western centric data collection from English language news reporting.

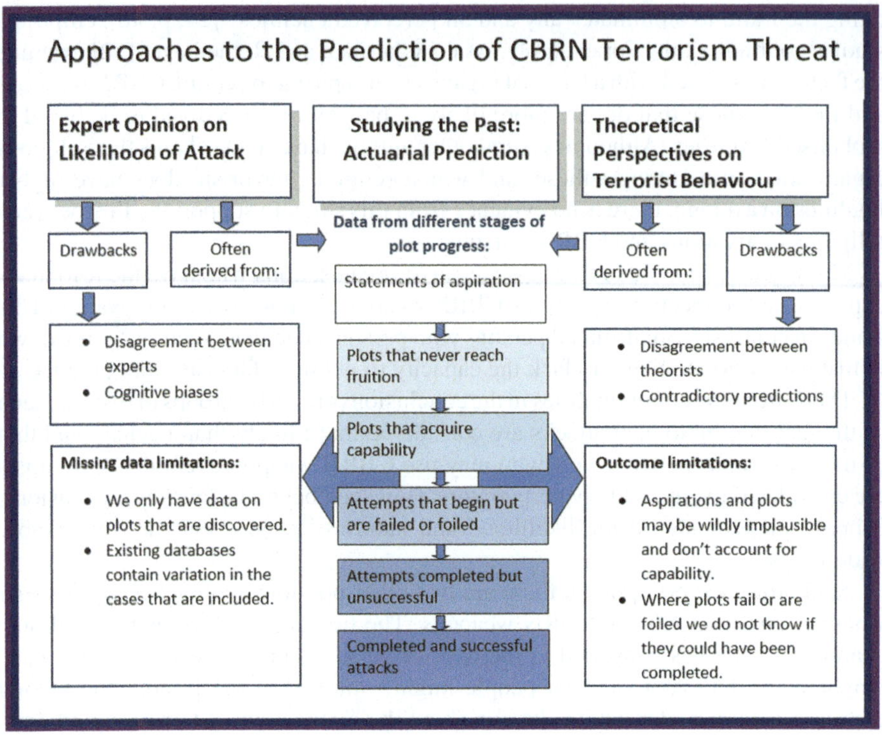

Fig. 1 Approaches to the prediction of CBRN terrorism threat (Credit: M. Wilson)

In order to understand the nature of potential threats in any area of criminal activity, psychologists and criminologists have frequently studied those attacks that have been carried out, or planned to have been carried out, in the past. Through empirical study of known or planned attacks it is possible to develop a framework for assessing the current threat in terms of understanding the likely perpetrators and potential targeting, and this can be used to help prevent attacks and to mitigate the effects of potential attacks should they occur. Although it is hard to predict the future from the past, systematic analysis can reveal how threats are evolving over time. Recent research priorities in security have included studies of behaviour change and establishing models of technology transfer that might "provide indications of when, where, how, and why a specific technology transfer might occur" [33, p. 28]. An analysis of this kind, examining CBRN attacks, would be a valuable contribution to the literature.

The study of past cases can give an indication of the range of modus operandi (MO) for such attacks. Studies of targeting in attacks involving explosive devices have shown that one of the dimensions that distinguishes between forms of attack is 'Proximity and Involvement'. This dimension distinguishes between the more remote and 'hands off' style modes of delivery such as postal, through to those such as suicide bombings, where the perpetrator is not only at the scene, but is also usually a victim [34]. It is likely that this dimension will also play an important role in distinguishing CBRN attacks, where degrees of physical involvement may be seen in the mode of delivery, ranging from hands-off methods such as postal delivery (as seen in the anthrax letters in 2001) and product tampering, to those where direct contact is needed between the victim and the perpetrator, such as the 2017 assassination of Kim Jong-nam.

In order to understand the range of terrorist use or attempted use of CBRN, an examination of less widely cited attacks may give us more insight into what has been attempted and against whom. With respect to attacks that have a terrorist motivation, the largest and most widely cited source is the Global Terrorism Database (GTD). GTD records all attacks by terrorist actors worldwide, and records, amongst many other variables, the nature of the weapons used in the attack. Despite being the most widely used database of terrorist action, this data set is not infallible. There will be unsuccessful cases, foiled cases and, by definition, non-terrorist cases that are omitted.

A search of this database shows that worldwide, between 2012 and 2017, four incidents were classified as using biological weapons. Three of these were carried out by anti-gun control extremists in the US in 2013 and one was carried out in Colombia by the Revolutionary Armed Forces of Colombia (FARC) in 2014. The US cases were all instances of letters containing ricin which were mailed to prominent persons on the same day and all were intercepted before they reached their intended victims. The FARC attack was coded as biological because the explosives also contained a cylinder of sand mixed with faeces. The device was discovered and diffused.

By comparison, those incidents classified as using chemical weapons were far more common with 172 identified in the same time period and occurring in countries worldwide. Attacks are coded as chemical with a greater frequency because of the relative ease of obtaining certain materials and include acid attacks, pepper spray and rat poison. Often, however, there is little detail provided in this database as to what the agent used actually was and they are often described simply as "poison", or "toxic gas". While these cases include contamination incidents such as poisoning water supplies, many cases involve explosives (rockets/mortars/IEDs) that are "laced" with toxic agents. There are no incidents recorded during this time period that were coded as having used nuclear or radiological weapons, perhaps reflecting the difficulty of obtaining such materials.

What is clear from this data search is that substances that are classified as Chemical or Biological for the purposes of the GTD, are not the same as those typically denoted by the term CBRN. This is an important point to note, since the GTD is widely used by both academic and security practitioners. Without careful examination, using these categories "off the shelf" would result in an inaccurate portrait of the nature and prevalence of these types of attack. This problem also relates to the research on product tampering, where biological matter or chemical substances are introduced into a product, but are usually very common products which are easy to obtain.

While the GTD is a very valuable resource for those who study terrorist attacks, it does not contain attacks that were planned and failed, or planned and foiled by the authorities before they came to fruition. It has been noted that failed and foiled CBRN attacks should not result in complacency over how hard they are to accomplish, but rather be seen as a sign of intention, which may one day result in a successful attack [25]. While the BAAD dataset does contain attempted CBRN attacks, it does not distinguish between those with intentions only and those that actually used such weapons. For attacks that went beyond planning, but still failed or were foiled, we can consult the Failed Foiled Completed and Successful (FFCS) database, which may give us a better understanding of what terrorist actors aimed to carry out and how, as well as how they came to fail [35]. While this database is an interesting resource, it only contains cases considered to have "jihadist" motivations. While this will contain cases where the perpetrators were in direct contact with terrorist organisations, it also contains lone actors who self-identified with these types of causes, while in reality have no contact with other terrorists at all.

The data set is further limited by date (1993–2017) and the location of the attack, containing only incidents that occurred in the US, Canada, Australia, New Zealand and the EU and NATO countries. Since we can only obtain data on those planned attacks that were discovered, naturally, there may be many other attacks that are planned but never occur or are identified. Finally, because the data is constructed from news sources, there is no independent verification of what the terrorists attempted and, often, the reports lack detail on the specific substances used, or indeed whether they were ever actually acquired. While the data set is limited in its scope, it does reveal more detail of the aspiration of these terrorist actors in relation to CBRN. There were 13 such attacks identified and none of them were successful, in that they did not harm any of the persons toward whom they were directed. However, the database also considers whether the attacks were completed, that is, whether the sequence of events was carried out even though no harm came to the targets. A case would be completed, for example, if a letter containing ricin was delivered to the recipient, even though they did not eventually open it or come to harm. Finally, the database also provides more detail on how the plots came to be failed or foiled.

In terms of their overall coding, the majority of the 13 planned attacks were chemical (10), with two cases coded as radiological and one biological. There were no attacks that were nuclear. The targets of these intended attacks were frequently governmental or military. For instance, a 2001 plot to attack the European parliament in Strasbourg with sarin gas was foiled. The reports do not specify how close the terrorists were to being able to carry out this attack. On 3 June 2003, six letters containing a mixture of phenarsazine chloride and hydrazine were sent to the Belgian Prime Minister's office as well as the British and Saudi Arabian embassies in Brussels. Others were sent to various Belgian ministries, the Brussels courthouse, an airport in the town of Oostende and the port authority of Antwerp.

Embassies were frequent targets in this sample, with US diplomats targeted in embassies in The Hague, Netherlands and Belgium. In 2001 a letter containing cyanide was sent to the US embassy in Wellington, New Zealand, and in 2003, a group known as September 11 claimed responsibility for the earlier action and sent further cyanide-laced letters to the US Embassy and the British and Australian High Commissions in Wellington and the New Zealand Herald Newspaper in Auckland. Aside from the symbolic government targets in these incidents there were other attempts to attack a more general civilian population. For example in the 2003 foiled plot to spread ricin on the Heathrow Express train in London, UK; the 2004 foiled "dirty bomb" plot aimed at various public targets in London; the 2002 plot to poison the water supply in Rome; and the 2003 failed attempt to release cyanide gas in the New York subway.

6 Specific Forms of Action Related to Terrorism

6.1 Assassinations

As mentioned earlier, it is difficult ever to classify assassinations as terrorism, as they do not typically fulfil the criteria required for the definition [8]. Nevertheless, assassinations are sometimes carried out by terrorist organisations, and although terrorist assassinations make up only a small proportion of assassinations in general, they deserve some consideration here. We have already established that few terrorist attacks use CBRN weapons, and since we can say the same for assassinations in general, then it seems reasonable to conclude that terrorist assassinations that use CBRN weapons will be very rare indeed [36].

CBRN assassinations spring to mind precisely because they are unusual, providing another example of the availability heuristic [28]. Such cases as the London assassinations of Georgi Markov in 1978 and Alexander Litvinenko in 2006, the Kuala Lumpur 2017 assassination of Kim Jong-nam, the 2018 attempted assassination of Sergei Skripal and the 2020 attempt against Alexei Navalny all gained extensive media coverage. All of these cases were confirmed or suspected to be carried out by government sponsored assassins, which accords with the previously cited argument that CBRN materials are challenging for non-state actors to obtain or to weaponize. This challenge, along with a number of other potential explanations, reviewed below, account for the low frequency of CBRN attacks associated with terrorism. However, in relation to assassinations, poisonings of every kind are low frequency because they involve a lot of planning and close access to the victim. For this reason, they are often carried out by or with the help of "insiders".

6.2 Malicious Contamination

A range of different threat actors use malicious contamination (of varying degrees of severity) for a variety of reasons. At one end of the scale, there are those with personal motivations such as financial extortion or revenge. Here we may see products being contaminated, and the manufacturers threatened that products will be released to the public unless a 'ransom' is paid. Cases of revenge tend to be either direct personal attacks using poisons against known others such as neighbours or work colleagues, or disgruntled employees, using contamination to damage a company's reputation or create financial damage to their current or former employer. At the political end of the spectrum, we also see contamination of food and water supplies, as well as other products such as medicines and cosmetics. While some of these political actors use contamination to place pressure on certain manufacturers to draw attention to causes such as animal rights, others intend their acts to do actual harm to large numbers of people. These acts of "food terrorism", in common with the definition of terrorism in general, are intended to draw attention to some political, social, or religious cause. However, in a minority of cases, targeting random people, purely for malicious reasons is motivation in itself [23, 37].

Interestingly, as Kilbane [38] highlights, for many contamination crimes, actual contamination is not necessary in order to achieve the perpetrators' objectives. For example, a credible claim to have successfully contaminated a product may be enough to use for extortion purposes against a company, or to cause economic or reputational damage. For this reason, Kilbane examines both genuine acts of contamination as well as hoaxes and false claims. Her results show that while CBRN contaminations are often claimed to have been achieved by the perpetrators, presumably because of the high levels of fear they can induce, they are rarely present in genuine contaminations.

6.3 Insider Attacks

Finally, mention should be made of research on insider threat. Over the last decade research on security threats has turned to the role played by so called "insiders", that is those who already have legitimate access to various information or procedures within an organization which they could use for malicious intent. The range of threats covered by this research is extensive, from those who wish to harm the organization itself, such as disgruntled employees, to those who wish to use their access to harm others in pursuance of wider goals, for instance a terrorist organization. It is not solely those who might infiltrate a workplace that are of concern, but also the vulnerability of legitimate staff to threats or coercion. Thus, it is thought that the threat is less likely to be associated with terrorist actors who attempt to join the organizations they wish to infiltrate, but rather from people recruited by them from employees already in place [29].

In relation to potential vulnerabilities and modes of attack, insiders have been flagged as a particular concern to the nuclear, chemical and biotech industries. There have always been fears that attacks against or sabotage by insiders at nuclear or other facilities could result in disastrous consequences. Indeed, there have been armed assaults, planted explosives, and cyber hacks detailed in the past, although it is fair to say that these are very few in number [39]. A number of authors have discussed the physical and other practical measures that could be put in place in order to deter or detect potential sabotage against nuclear facilities [see for example, 40].

Bunn and Bunn [41] point out that although material in international transportation is the only area in which there are binding international security requirements, concerns remain over the susceptibility of nuclear material in transit to theft or other tampering. Similarly, Greenberg, Chalk and Willis [42] list nuclear material in transit as one of the key potential future vulnerabilities in their analysis of terrorist threat. Bunn, et al., [39] suggest that the real barrier to strengthening nuclear security is the belief that nuclear terrorism is extremely unlikely, which they put down to a lack of knowledge of the capability of potential thieves worldwide. They claim that the most convincing evidence of the threat comes not from the examples given of terrorist organisations who have stated their *interest* in acquiring nuclear weapons, but rather in the evident failures in nuclear security of the past. Between 1992 and 2019 they identify more than 20 cases where small quantities of plutonium or highly enriched uranium (HEU) have been stolen and recovered.

7 Theoretical Perspectives

The second approach to assessing the likelihood and nature of potential terrorist CBRN attacks is to consider the theory of terrorist behaviour to try to predict how terrorist actors would behave and why. This analysis identifies a number of factors that could predict a positive payoff to terrorists in using CBRN weapons, balanced against the potential negative consequences.

7.1 Capability

While we have already considered the technical knowledge and resource issues around accessing and weaponizing CBRN, there is a further capability issue we need to consider: the role of human actors. Campbell and Murdie [43] highlight an extraordinarily important link in the chain of events that must be in place in order for a CBRN attack to take place. Following the pyramid model of radicalization [44, 45] it is well established that those terrorist actors who are actively engaged in violence are backed up by a larger base of supporters and a larger still base of sympathizers, that the actors rely on for resources such as safe houses, etc. The development of any terrorist plot requires a number of people who have knowledge, or suspicions, *not to report* [35], and this applies even more so to a plot involving CBRN weapons. The plan for such an attack would need people to remain silent, from the immediate population who surround the terrorist actors themselves through to the personnel who must be drawn upon in order to obtain the materials required [25]. The possible consequences of a CBRN attack may place more pressure on people to report than would an attack using traditional weapons because of the potential for many indiscriminate victims. In this respect, Campbell and Murdie [43], along with a number of other researchers [e.g., 2, 46], highlight the relationship between corruption, repressive regimes and potential CBRN terrorism. It is likely that widespread corruption would facilitate those wishing to obtain access to CBRN materials. In addition, repressive regimes are often characterised by low levels of trust in the government, and they therefore make public reporting of suspicious behaviour less likely.

7.2 Desirability

Returning to the definition of terrorism, the aim of true terrorist incidents is to create fear in the general population, and to use that fear to put pressure on an authority (most often a government) in order to obtain a concession of some sort, usually political or related change. CBRN attacks (or the threat of them) are thought to be very effective in generating fear, and some have argued that the invisibility of the agents and the uncertainty over exposure could be more *psychologically* threatening than traditional attacks using guns and bombs [47, 48]. One of the arguments that terrorists would move to CBRN attacks in the future is that because terrorism relies on fear, and fear is heightened by novel threats, terrorists may resort to such methods in order to increase the fear factor, which has become too commonplace through ordinary means.

However, the question that remains is whether the extra effort required to obtain and design an effective delivery system to weaponize CBRN is worth it for the return. To have a positive return, one has to believe that terrorist actors want a large number of people dead, and many terrorism experts do not believe this to be true. This position is summarised by the well-known quote that terrorists 'want a lot of people watching, not a lot of people dead' [49, p. 583]. However, in the immediate aftermath of 9/11, a number of terrorism researchers came to believe that terrorism was undergoing a fundamental change, involving less rational and political action and a greater emphasis on mass casualty attacks. Supporters of the "new terrorism" theory cite the increasing number of casualties associated with terrorist action, and while this in itself is not disputed, other theorists remind us that there are other possible explanations [e.g., 50] and that the small number of high profile mass casualty attacks that affect yearly averages do not reflect on terrorism as a whole, which is characterised by thousands of smaller scale incidents, many of which have no casualties at all. Some authors have therefore argued that we must place the fear of terrorists using WMDs into a wider context [8] and note that 'innovative use of old techniques can still result in many casualties [51, p. 52]. Countering theories of "new terrorism", Crenshaw [51] concludes that motivations have not changed and that it is not necessary to throw away old knowledge in favour of the new [52]. Likewise, Ackerman and Asal [53] also conclude that mass casualty terrorism is not a new phenomenon and most attacks are conducted using bombs and guns.

In relation to casualties then, the terrorism research community remain divided in their opinion of whether the mass casualties of CBRN would be desirable to terrorist actors or not. What is less ambiguous is the role of targeting. A mass casualty attack using CBRN would in all likelihood mean very imprecise targeting and therefore risk of injury, illness or death to members of the terrorists' own constituency. For this reason, Ackerman and Snyder [24] conclude that CBRN weapons are highly unlikely to be used by terrorist groups in the Middle East, even if they have the capability. However, not all CBRN attacks are mass casualty, and so precisely targeted, low casualty incidents, such as assassinations are arguably more likely than large scale terrorist attacks. Finally, threats and hoaxes are still defined as terrorism and can create fear and disruption to publicise the terrorists' cause, without requiring the resources and avoiding the inevitable negative consequences.

As mentioned earlier in the chapter, what is important to terrorists is not necessarily the number dead per se, but the publicity (good and bad) and so the target, the victims and the public reaction are key considerations in defining the "success" of terrorist action. There is a good deal of theory to draw on from Terrorism Studies in relation to the way that the media portrays terrorism, and the apparent preference for more unusual or sensational acts of violence to be covered in the media than those that use more mainstream methods. It is therefore argued that the media mediates between terrorism and public fear, acting as an amplifier for the message the terrorists aim to communicate. Therefore, some academic authors have argued that terrorists are being pressed into taking more extreme action in order to gain the publicity they need.

However, while increasing novelty and casualties would predict increased publicity, not all publicity is good publicity. As seen in relation to claims of responsibility, there are certain types of attack that may be seen to have "crossed a line", even for terrorists. The counterargument in relation to publicity and CBRN is that such attacks, especially indiscriminate ones, would be likely to be met with a poor reaction from the public. As discussed at the start of this chapter, many terrorist attacks go unclaimed, and one of the reasons for this is when the attack results in a backlash, especially from the terrorists' own constituency. For example, it is widely accepted that the public backlash against ETA following attacks which injured and killed Spanish civilians contributed to the demise of the organisation. Thus, on a purely theoretical level, one can argue that using CBRN in a mass casualty capacity has the potential to be badly received and would therefore be avoided.

7.3 *Retaliation*

The final area where we might find some clues as to why terrorist actors would aspire to use or avoid CBRN attacks is that of the potential consequences for themselves. Here again, we can take some direction from the research on claims of responsibility.

It could be argued that terrorists may see obtaining CBRN weapons as giving them greater bargaining power because of the potential threat they now hold. However, in relation to terrorist attacks in general, Hoffman [20] and Rapaport [19] have both drawn attention to fears of reprisals against the terrorists themselves, and/or the countries that protect them. In this respect, Campbell and Murdie [43] predict that a CBRN attack would likely elicit an extreme response from the targeted government. While this can be seen as a clear deterrent, we also need to consider a counter possibility.

A number of terrorism theorists have stressed the importance that government actors *do not* engage in retaliatory action, because such action leads to greater support for the terrorists' cause (e.g., 54–56]. For example, in setting out the theory of 'jujitsu politics', McCauley [56] explains that the *very aim* of some terrorist action is to provoke a government into a heavy-handed response that affects the ordinary citizens who surround a terrorist organisation. This attack by the government then leads more of the population to see the 'enemy' government in a bad light and subsequently increases terrorist recruitment to the cause. It could therefore be argued that far from being a fear, retaliation would be welcomed as it contributes to the terrorists' longer term aims.

In conclusion, based on terrorism theory, while there are positive motivations involving fear and displays of power in having the public, the authorities and rival organisations *think* they have the capability for a CBRN attack, most terrorist organisations do not, and if they did they would be unlikely to use it, especially in relation to mass casualty terrorism.

8 Conclusions

This chapter has reviewed the literature regarding the potential for terrorist actors to engage in CBRN attacks in the future; considering evidence from studies of expert opinion, past incidents and the theoretical arguments derived from Terrorism Studies. All of these methods point to the unlikelihood of terrorist actors using CBRN weapons, especially in relation to mass casualty terrorism. The conclusion from all three approaches is that while there may be a very small number of terrorist actors who aspire to CBRN attacks, it is a rare aspiration and most acts of terrorism remain conducted using the traditional methods, most often bombs and guns.

While the evidence suggests that terrorist actors are unlikely to use CBRN weapons, it is important to stress that there are many other actors who might, including state actors and individual extremists with non-terrorist agendas. These low-risk high consequences attacks still require serious consideration in relation to the motivations and behavioural precursors that might predict and therefore prevent an attack in the future.

References

1. Jackson BA, Frelinger DR (2008) Rifling through the terrorists' arsenal: exploring groups' weapon choices and technology strategies. Studies Conflict Terrorism 31(7):583–604
2. Ivanova K, Sandler T (2007) CBRN attack perpetrators: an empirical study. Foreign Policy Anal 3:273–294
3. Ackerman GA (2018) Chemical, biological, radiological and nuclear (CBRN) Terrorism. Routledge handbook of terrorism and counterterrorism. Routledge, pp 240–252
4. Schmid A, Jongman A (1988) Political terrorism: a new guide to actors, authors, concepts, data bases, theories and literature. Transaction Books, New Brunswick, NJ
5. Schmid AP (1993) Defining terrorism: the response problem as a definition problem. In: Schmid AP, Crelinsten RD (Eds) Western responses to Terrorism. Frank Cass, London
6. Beck CJ, Miner E (2013) Who gets designated a terrorist and why? Soc Forces 9(13):837–872
7. Logan B (1993) Product tampering crime: a review. J Forensic Sci 38(4):918–927
8. Schmid AP (2000) Terrorism and the use of weapons of mass destruction: from where the risk? In: Taylor M, Horgan J (Eds) The future of terrorism. Frank Cass, London
9. Wilson MA, Lemanski L (2010) The forensic psychology of terrorism. In: Adler J, Gray J (eds) Forensic psychology; concepts, debates and practice. Willan Publishing, Second Edition, pp 245–263
10. Enders W, Sandler T, Parise GF (1992) An econometric analysis of the impact of terrorism on tourism. Kyklos 45(4):531–554
11. Yuchtman-Yaar E, Hermann T (1998) The latitude of acceptance: Israeli attitudes toward political protest before and after the assassination of Yitzhak Rabin. J Conflict Resolut 42:721–743
12. Iqbal Z, Zorn C (2008) The political consequences of assassination. J Conflict Resolut 52(3):385–400
13. Bruck T, Wickstrom B (2004) The economic consequences of terror: guest editors' introduction. Eur J Polit Econ 20:293–300
14. Jones BF, Olken BA (2009) Hit or miss? The effect of assassinations on institutions and war. Am Econ J Macroecon 1(2):55–87
15. Borum R (2004) Psychology of terrorism. University of South Florida, Tampa FL
16. Kearns E, Conlon B, Young J (2014) Lying about terrorism. Studies Conflict Terrorism 37(5):422–439
17. Bradford E, Wilson MA (2013) When terrorists target schools: an exploratory analysis of attacks on educational institutions. J Police Criminal Psychol (Special Issue) Terrorism Psychol Theory Appl 28:127–138
18. Bradford EJ, Wilson MA (2015) Tactical targeting: a theoretical perspective of terrorist attacks on schools. In: Stedmon A, Lawson G (Eds) Hostile intent and counterterrorism: human factors theory and application, pp 233–244
19. Rapoport D (1997) To claim or not to claim; that is the question—always! Terrorism Political Violence 9(1):11–17
20. Hoffman B (1997) Why terrorists don't claim credit. Terrorism Political Violence 9(1):1–6
21. Hoffman A (2010) Voice and silence: why groups take credit for acts of terror. J Peace Res 47(5):615–626
22. Kearns E (2019) When to take credit for terrorism? A cross-national examination of claims and attributions. Terrorism Political Violence Online first https://doi.org/10.1080/09546553.2018.1540982
23. Wilson MA, Kilbane S (2016) Criminal poisoning and product tampering: toward an operational definition of malicious contamination. Deviant Behav 38(10):1141–1159
24. Ackerman G, Snyder L (2002) Would they if they could? Bull Atomic Scientists
25. Danzig R, Sageman M, Leighton, T et al (2012) Aum Shinrikyo: insights into how terrorists develop biological and chemical weapons. Center for a New American Security
26. Boddie C, Watson M, Ackerman G, Gronvall GK (2015) Assessing the bioweapons threat. Science

27. Koblentz GD (2011) Predicting peril or the peril of prediction? Assessing the risk of CBRN terrorism. Terrorism Political Violence 23(4):501–520
28. Tversky A, Kahneman D (1974) Judgement under uncertainty: heuristics and biases. Science 125:1124–1131
29. Martin P (2019) The rules of security: staying safe in a risky world. Oxford University Press, Oxford
30. Koehler D, Popella P (2020) Mapping Far-right chemical, biological, radiological, and nuclear (CBRN) terrorism efforts in the west: characteristics of plots and perpetrators for future threat assessment. Terrorism Political Violence 32(8):1666–1690
31. Stenersen A (2009) Al-Qaida's quest for weapons of mass destruction: the history behind the hype. VDM Verlag, Saarbrücken
32. Asal VH, Ackerman GA, Rethemeyer RK (2012) Connections can be toxic: terrorist organizational factors and the pursuit of CBRN weapons. Studies Conflict Terrorism 35(3):229–254
33. Earnhardt R, Ackerman G (2019) Modelling terrorist technology transfer. On the horizon: security challenges at the nexus of state and non-state actors and emerging/disruptive technologies (April). https://nsiteam.com/social/wp-content/uploads/2019/04/DoD_DHS-On-the-Horizon-White-Paper-_FINAL.pdf
34. Wilson MA, Lemanski L (2010) Key dimensions in understanding terrorist bomb attacks. Final report to DHS. START
35. Crenshaw M, Dahl E, Wilson M (2017) Comparing failed, foiled, completed and successful terrorist attacks: final report year 5. Report to the office of university programs, science and technology directorate, U.S. Department of Homeland Security. College Park, MD: START
36. Wilson MA, Scholes A (in review) The role of organization in political assassinations
37. Kilbane S, Wilson MA (2021) A definitional framework of poisoning crimes. In: Conyers A, Calhoun TC (Eds) Deviance today, 2nd edn, Abingdon: Routledge
38. Kilbane SC (2018) Agent Selection and threat actualization in contamination cases: predicting action from perpetrator behaviour. J Threat Assess Manag 5(3):173–187
39. Bunn M, Roth N, Tobey WH (2019) Combating complacency about nuclear terrorism. Policy briefing: harvard Kennedy school, Belfer Center for Science and International Affairs
40. Chen S, Demachi K (2019) Proposal of an insider sabotage detection method for nuclear security using deep learning. J Nucl Sci Technol 56(7):599–607
41. Bunn M, Bunn N (2001) Nuclear theft & sabotage priorities for reducing new threats reducing the threat of nuclear theft & sabotage. Paper presented at the IAEA international symposium on safeguards
42. Greenberg MD, Chalk P, Willis HH (2006) Maritime terrorism: risk & liability. RAND: Santa Monica, Arlington, Pittsburgh
43. Campbell B, Murdie A (2018) keep the informants talking: the pursuit and use of CBRN weapons by terrorist organisations. Studies Conflict Terrorism. https://doi.org/10.1080/1057610X.2018.1531517
44. McCauley C, Moskalenko S (2008) Mechanisms of political radicalization: pathways toward terrorism. Terrorism Political Violence 20:405–433
45. McCauley C, Moskalenko S (2017) Understanding political radicalization: the two-pyramids model. Am Psychol 72(3):205–216
46. Early B, Furhmann M, Li Q (2013) Atoms for terror? nuclear programs and non-catastrophic nuclear and radiological terrorism. British J Political Sci 43(4):915–936
47. Rogers MB, Amlôt R, Rubin GJ et al (2007) Mediating the social and psychological impacts of terrorist attacks: the role of risk perception and risk communication. Int Rev Psychiatry 19(3):279–288
48. Rubin GJ, Chowdhury AK, Amlôt R (2012) How to communicate with the public about chemical. Biological, radiological or nuclear terrorism: a systematic review of the literature, biosecurity and bioterrorism: biodefense strategy. Practice Sci 10(4):383–395
49. Jenkins BM (1987) The future course of international terrorism. In: Wilkinson P, Stewart AM (Eds) Contemporary research on terrorism. Aberdeen University Press, Aberdeen

50. Asal V, Blum A (2005) Holy terror and mass killings? Reexamining the motivations and methods of mass casualty terrorists. Int Studies Rev 7(1):153–155
51. Crenshaw M (2006) Have motivations for terrorism changed? In Victoroff J (Ed) Tangled roots: social and psychological factors in the genesis of terrorism. IOS Press, Oxford
52. Wilson MA (in press) Terrorism research: current issues and debates. In: Brown J, Horvath M (Eds) The cambridge handbook of forensic psychology. 2nd Edn. Cambridge University Press
53. Ackerman G, Asal V (2011) Understanding and combating mass casualty terrorism. University of Maryland, USA, National Consortium for the Study of Terrorism and Responses to Terrorism
54. Cronin AK (2007) Chapter one: the strategies of terrorism. Adelphi Papers 47(394):11–22
55. Kydd AH, Walter BF (2006) The strategies of terrorism. Int Secur 31(1):49–79
56. McCauley C (2006) Jujitsu politics: terrorism and response to terrorism. In: Kimmel PR, Stout CE (eds) Collateral damage: the psychological consequences of America's war on terrorism. Praeger, Westport, CT, pp 45–65

Human Factors and Societal Aspects in Future CBRNE Incidents

K. Brattekås

Abstract CBRNE agents may pose a potent physical threat but may equally–or even more so–create significant fear and public distress. This chapter explores how the public can react to CBRNE, how states may handle incidents and how technology may be utilised to counter adverse societal effects.

1 Introduction

As worldwide travel and freight becomes increasingly common, the global population grows and the climate changes, there is nothing to suggest that natural or accidental CBRNE incidents will become less frequent. According to a NATO-study [26], the CBRNE security threat is likely not decreasing either:

> Due to the increasing number of terrorist attacks, maintaining security has become a significant concern in recent years and is likely to remain so because state and non-state actors have more access to lethal technologies. Use of … Chemical, Biological, Radiological, Nuclear (CBRN) or toxic industrial materials could give terrorist and radical groups an increasing ability to stage technologically advanced attacks that could create overwhelming effects. Additionally, possession of significant employable weapons of mass destruction (WMD) capacity by 'Strategic States' would result in a vast potential for harm, if they succumb to sudden catastrophic instability or failure [26].

States should thus be prepared for future CBRNE incidents—small and large scale—such as new pandemics, industrial accidents, terrorist attacks, acts of war and sabotage.

CBRNE incidents, whether they are deliberate or accidentally caused, have direct and indirect consequences for people and society. The impacts will likely be significant in the short and long term. In the planning for, and management of, CBRNE

K. Brattekås (✉)
Norwegian Defence Research Establishment (Forsvarets Forskningsinstitutt (FFI)), P.O. Box 25, NO2027 Kjeller, Norway
e-mail: kjerbra@gmail.com

P.D.E. Biggins and D. Chana (eds.), *CBRNE: Challenges in the 21st Century*, Advanced Sciences and Technologies for Security Applications,
https://doi.org/10.1007/978-3-031-17374-5_8

incidents, the population's reactions, and their potential for being a resource—before, during and after—should be taken into consideration.

This chapter addresses research on human response, with emphasis on the fact that CBRNE incidents will likely be experienced in different ways to more frequent or likely risks. For example, CBRNE is often characterised as a *low-probability, high-impact* risk, which often involves agents invisible to the naked eye. Thus, CBRNE incidents may require customised preparedness and response from both the managing bodies and the public.

Some important success factors for resilience in society when CBRNE incidents occur are to be able to timely and efficiently address the populations affected and create, or rather earn, trust in order for people to comply with official advice. The main recommendation from this chapter is to take into consideration human and societal factors in planning, preparedness, response and long-term management of CBRNE.

A key concept in this chapter is *society*, yet there is no space to explore the concept in full here. Sovereign states (referred to as 'states') are the entities that will be addressed here as representatives for societies. States in this context refer to the legal entities recognised by the United Nations. This is, of course, a gross simplification, as states as legal entities may be equal, but are vastly different in terms of size, economy, type of government, politics etc. The concepts of state, government and country are also often used interchangeably and without consistency [46]. The relevance here is that all states to some degree are responsible for the safety and security of their population. This is part of the social contract[1] between states and citizens, and one of the main pillars of society as we know it today.

This chapter initially focuses on how the research suggests engaging the public to help themselves and others and discusses dilemmas of compliance. However, it is ultimately states, i.e., the leaders, governments, and central managements that should be held responsible for crisis preparedness and response.

- As food for thought before delving into this chapter, consider the recent (2020–22) coronavirus (COVID-19) pandemic. Even though the jury is still out on what is or would have been the best response to the pandemic, we have seen that different states went for different strategies in the crisis management. Were any states adequately prepared for a pandemic? Did states provide sufficient information for their citizens? Did states help each other and exchange critical information about the virus and vaccine development? These are questions that should be addressed in the evaluation of the crisis.

---

[1] Social contract theory can be dated back to the 16- and 1700s, with influential thinkers such as Thomas Hobbes, Jean-Jacques Rosseau and John Locke in the forefront. In some form or another, all three argued that the state of nature and humans needed to be regulated in a form of social contract where humans gave up some of their liberties, or even rights, to the state in exchange for protection of their lives and property.

This chapter starts with exploring the research on what we know about people's perceptions and actions in CBRNE incidents and the main recommendations for engaging the public. The chapter goes on to discuss how society may develop in the future, and how technology can provide both solutions and add dilemmas, before summing up the key points for managing future CBRNE threats in a societal context.

2 Perceptions of CBRNE

CBRNE agents are different in nature, yet what they have in common is the potential to create significant fear. People may fear for the adverse effects of radiation, chemicals and biological agents that can be hard to detect with our senses. Both natural and accidental CBRNE incidents can potentially create mass fear in the population, but the thought of adversaries using these means to create damage may well be the most potent threat.

In short: CBRNE agents can, understandably, be perceived as scary. This is, in essence, the starting point for how people perceive the unfamiliar and often invisible threat. Several studies have investigated how people react to CBRNE incidents and how communication about such incidents should be conducted. The subjects have been explored from different angles, but the topics seem to remain understudied [20]. This could be due to the low likelihood—even if the consequences are massive—or the complexity of the field(s). Human and societal factors in CBRNE is not *one* field of study, it is several: communication studies, psychology, social psychology, political science, technical CBRNE knowledge, security studies or medicine only to mention a few. Thus, it remains a vast area of study with several directions unexplored.

In this section, some literature on *what we do know* is summarised in order to light the way on how states could proceed to address people's concerns.

2.1 How Do People Behave?

While there is a large and established literature on the fear generated by both natural disasters and various forms of malicious attacks against the population, the use of CBRNE agents has the potential to create a qualitatively different kind of fear. As mentioned, this is mainly due to their unfamiliar nature and potential dire consequences. People's perceptions, societal-, fear- and dread effects are crucial to address when planning for and managing incidents [33, 34, 43].

In some cases, how people react, and act, can determine how a CBRNE incident plays out. The radiation incident in Goiânia, Brazil serves as an example: two men entered an abandoned hospital site in 1987, where they found a container with blue, iridescent powder. They took the container containing the powder; this resulted in several people handling the powder which emitted a glow. Unbeknownst to them, the powder was a highly radioactive source. This in turn led to four deaths, several people with radiation damage and contamination, and ultimately an extremely costly clean-up of the site and surroundings [17]. Furthermore, the people of Goiânia were subjected to harsh stigmatisation. The surrounding areas shunned anything to come out of Goiânia, including people, vehicles and goods, which made conditions extremely poor for the years to come.

Stigmatisation is an issue that can occur if people are perceived as *contaminated*. It can be regarded as a social process, where the unfortunate endpoint could be social exclusion and discrimination of a group of people [31]. Some typical examples of stigmatised groups are people with disabilities, people with obesity, and the HIV-positive—especially in some parts of the world. Due to their perceived conditions, they can be marked as people with low 'social value', and as a result their 'social value' further declines [31].

Stigmatisation can be the result of escalating social processes and can be driven by such things as media reporting or rumours. For example, during and after the so-called *Spanish flu* (1918–20), Spaniards were stigmatised even though the disease most likely did not originate in Spain. They were, however, the first to report it. Furthermore, during the 2009 H1N1 pandemic, or the so-called *swine flu*, Mexicans were stigmatised due to allegations that the virus originated in Mexican pig farms [23].

In the vast majority of cases, stigmatised groups are victims of ignorance. This was also the case in Goiânia, where people and goods were perceived to be 'radioactive', and hence untouchable, by outsiders. This was of course not the case, yet the fear of invisible radiation accelerated by misinformation had massive impacts for the area. In economic and social terms, the stigmatisation effects were probably much larger than the radiation incident itself. Where stigmatisation occurs, including in incidents involving CBRNE agents, there is a need for information campaigns to address ignorance and fear.

Fear and anxiety created by CBRNE agents may even cause physical symptoms from fear of exposure for some people. Individuals who are worried that they have been impacted, but who are not actually contaminated, are often referred to as the 'worried well' [12, 36, 45]. Stigma and fear can also impede management efforts [1]. Thus, CBRNE incidents may also be described as *dread risks*, which means that invisible agents may cause more fear than the damage they actually do [9]. However, the fear and anxiety induced in turn can create a larger problem.

An example of this is the 2001 *anthrax letters* posted to officials in the USA. In chap. 1, the letters killed and hurt only a minimal number of people, but initiated programmes and measures with great costs for society [40].

Other researchers claim, based on their studies, that we do not give the public enough credit for their resilience, and that the public can be utilised as an ally in emergencies rather than to be assumed passive or paralysed when frightened. Furthermore, studies have shown that panic is in fact not as widespread as many assume, and that people tend to act rationally based on the information they have [8, 29, 36, 45]. This means that people are likely to follow official advice as long as it seems rational to them based on what they know. However, as discussed later in this chapter and in other chapters, people's perceptions can be distorted, and background knowledge varies greatly. Ensuring that people get access to the right type of information and advice on what to do is a theme that is addressed later in this chapter.

It should be considered that even if research shows that most people behave in an orderly fashion immediately after an incident, attitudes may change in the extended period after [36]. Long term management is aided by openness and fair treatment of the affected in the aftermath of an incident, as well as thorough follow up. One prediction is that acute effects after a CBRNE attack could be less than feared, and that the long-term effects will be more difficult to manage than imagined [45].

According to literature, the general public's understanding of official CBRNE strategies for management and prevention is exceptionally low recommendations provided by official statements regarding protection and prevention tend to be misunderstood and are often perceived as confusing or complex [11].

Krieger et al. [20] identify four factor sets as drivers for public response in CBRN events. These are:

- "[T]hreat perceptions and risk as a 'feeling'" i.e., that the way people perceive risk will influence their behaviour.
- "[P]erceptions of responders, trust and transparency", which means that different levels of trust impact people's compliance in adhering to e.g., responders and official advice.
- "[P]erceived efficacy and costs of response measures", including that people's own perception of how well they will be able to cope with an incident, the measures taken and the cost it will have will influence their behaviour.
- "[Q]uality of risk and crisis communication", which needs to be timely, open and adapted to the target group.

The four factors may be overlapping, and even competing, and will vary in influence depending on such things as social factors, type of incident and response [20].

Another study showed that people's willingness to receive information up front on preparedness depended on aspects such as prior knowledge, demographics and psychosocial factors. Furthermore, elements that could increase compliance with official advice in case of an incident included "trust, provision of information, emotional responses; efficacy and relationships" [45].

In other words, there is no *one size fits all* solution to address the public or predict their responses in case of a CBRNE incident. However, to apply customised approaches for different population groups is recommended.

2.2 Developing Trust and Communicating with the Public

To anticipate and direct people's response to CBRNE incidents as adequately as possible, then, there are several important considerations that should be made. Communication and trust stand out as two of the most important factors to engage the public. Communication is essential to put into action any response or strategy concerning CBRNE. Trust is essential for the communication to be successful.

As we have seen above, several factors influence responses, and as people are different—so will their actions be. Compliance can be particularly important for management strategies to succeed, e.g., that people quarantine if they are ill or evacuate a contaminated area in an orderly fashion. Thus, communication with the public requires customisation, but most of all the advice needs to come from trusted sources.

- For example, during the COVID-19 pandemic, it has been revealed that some governments and government representatives initially downplayed how serious the threat from the virus was in order not to create public distress. How successful do you think this approach has been? Were there any good reasons for not disclosing the severity of the threat?

There is a real demand from the public for accurate and trustworthy information when incidents happen. There is seldom a feeling that authorities are sharing too much information to the public in the aftermath. When it comes to informing the public, withholding information can be more damaging than stating openly that risks, threats, and incidents are unclear or unknown [15, 45].

Involving experts in communicating CBRNE threats to the public is another essential for distribution of factual information. However, it is also important to communicate expert knowledge in an understandable way based on how "lay people" will perceive it and addressing their actual concerns and grievances. Initial communications should be short and concise, and appeal to people's common sense. More detailed and complex information should follow in order to make people adhere to the advice [3, 21, 22, 38]. Remembering the social context of those the communication is meant to reach is recommended, as well as bearing in mind how official advice on risks can be influenced by politics, economy, values or assumptions [41].

Understanding and addressing differing cultural views may be of significant importance in this regard. Cultural and social differences may affect how people respond (or not respond) to information. In cases where responders or authorities fail to take this into account, it could have several adverse effects, including noncompliance with official advice or stigmatisation of groups that do not follow this advice due to misunderstandings.

2.3 The Trust Element

As mentioned initially, *trust* is thus a vital factor when dealing with communication of CBRNE risks and incidents. Trust is first and foremost built by sharing honest, clear and consistent information and discussing risks and concerns with the public [8, 10, 24, 32, 39]. Trust must be built and earned over time. Maintaining it is equally important, as trust is fragile and can change rapidly. The erosion of trust can be gradual or very rapid, but once lost or eroded it is difficult to get back.

The public expectation is that authorities should deliver trustworthy information. If authorities withhold information, this is alarming to the public. This means that their need for information is not met, and the official advice can be perceived as shallow and not credible. Information should be communicated by trusted spokespeople, and trust between the media and authorities is a great advantage in such incidents [14, 41].

However, it should also be mentioned that crisis communication can have several adverse effects. Public authorities can rarely be held accountable for their emergency communication in the aftermath of an incident in case the information was wrong or harmful. "The result is that public authorities largely have a free reign to communicate how they wish and do not have to comply with a range of requirements" [30]. This aspect underlines how important it is to ensure trustworthy information in an appropriate form.

It can be a demanding task to build and maintain trust and communicate risks to the public at the same time. If for example public health officials raise the alarm or initiate measures for emerging disease threats, they may be perceived as they are 'crying wolf' and spreading fear. On the other hand, in the case of an epidemic or pandemic, no preparation is considered good enough in retrospect. It is recommended that public health officials build a "surge capacity for public trust". This means building a viable health system that takes proper care of the population in regular times in order for the population to trust them in emergencies [1].

Developing trust may of course seem easier on paper than it is in reality. In this day and age, where people may be exposed to false information, and may not take official advice into account, trust becomes even more vital to build. Combatting false information, or 'fake news', promoting fact checks and factual information, and educating people on how to spot what is trustworthy or not may provide a solution. However, to succeed with these strategies presents a significant challenge.

It is also recognised that one of the precursors to the development of conspiracy theories is when there are "gaps" in the information provided to the public. These gaps may be filled with seemingly plausible alternative explanations and are often perceived as an attempt by the authorities to "cover up" some kind of more sinister circumstance [13]. Since it is thought that belief in conspiracy theories is not just correlated with low trust in government but may also play a causal role in creating mistrust, then withholding information in the event of a CBRNE incident may serve to exacerbate any problems that already exist in relation to the affected community.

Vaccine hesitancy, or "anti-vaxxers", are an example of why trust and communication are crucial elements to combat biological threat agents. Vaccine hesitancy is an issue that is becoming more frequent in the Western world than before. Whether due to scepticism for side effects, contents of the vaccines, the belief that disease is a natural part of building immunisation or other reasons, anti-vaccine movements have caused diseases like measles to make their way back into the spotlight.

The incident involving Andrew Wakefield's early 2000s fraudulent research led to some of the vaccine hesitancy we still see today. His claims that the triple MMR vaccine could cause cognitive and physical effects in children spurred a massive uproar against the vaccine. The following media coverage when the research was proven false has not succeeded in convincing many concerned parents otherwise [5]. Most of the information that vaccine hesitancy is built on comes from unreliable online sources and are not fact based [18]. Some governments (like Italy) have chosen to enforce mandatory vaccinations, whereas others uphold the right to choose whether to immunise.

- What does the vaccine hesitancy tell us about trust in governments? And should health officials and state representatives take some of the blame for not discussing vaccine programs more openly?

Summing up some of the main findings in the studies in the area, or in other words *what we do know*:

- CBRNE has the potential to cause more dread or fear than the damage caused by the actual agents. However, in acute incidents, panic is rare, and people tend to be more resilient than expected.
- People will likely follow advice if they think it is the rational thing to do based on their background knowledge (such as their risk perception, level of trust in the managing bodies, what is expected of them and which information they receive).
- Having trust in the agencies responsible for CBRNE management (often state authorities) is of essence for people to comply with official advice. Trust must be earned and is difficult to win back if you lose it.

As stated in most of the studies that do exist on these topics, further research is required. However, based on what we do know, it is safe to recommend that trusted and competent officials should put forth facts about a risk or threat in an understandable way, people should not be underestimated as assets in an emergency, messages should be adjusted and customised to different population groups, and as much as possible of the communication strategies should be planned ahead.

Furthermore, it should be anticipated in the future that the population needs to be an ally before, during and after a CBRNE incident. People should be involved as part of the solution to prevent incidents from happening by being vigilant, and helping themselves and others, if incidents should happen. In order to gain the population's help, as an asset, technological solutions may prove helpful and efficient, but do not come without dilemmas.

3 CBRNE in the Future Society

As we have seen, CBRNE constitutes a complex threat, which can occur on its own, or in combination with other means or events. Technological and societal development *will* impact the CBRNE threat in the future—probably in both positive and negative ways. As we progress, we may be able to distinguish CBRNE threats sooner, or even prevent threats altogether with new and/or better means. However, progress can also bring about new risks and threats, such as more complex industrial systems prone to natural events or human error, or by technology lowering barriers for acquiring CBRNE agents with malicious intent. More importantly in this context, people's perceptions may be distorted due to the new arenas for interactions and information sharing made possible by technology.

Even though predicting the future of development in society or how the CBRNE field will develop is impossible to do in detail, existing and emerging trends provide some indications. Anticipated future conflict developments and CBRNE trends are also described in Chap. 1. Several studies and actors attempt to indicate future global security trends [4, 25, 26].These studies show that there is a general tendency towards enhanced human security and global equality.[2] This means that larger portions of the global population's living conditions, such as income, accommodation, education, and health care get better and better.

However, for the political landscape regionally and globally, some states show signs of becoming more authoritarian and tensions are predicted to arise and increase between traditional and liberal values, as well as inequalities within states, which seem to be on the rise [4, 25, 26].The outlets for this tension are hard to predict, but should the result be conflict, we may see more use of CBRNE materials in conflict zones. Some of the established norms that were largely driven forth by liberal states, such as zero tolerance for the use of chemical and biological weapons, could potentially be weakened if the power balance shifts [2].

Furthermore, it is predicted that the risk of inter-state conflict will remain low despite tensions rising, but sub-threshold conflicts, often termed "hybrid" or "grey zone" may become more common. There are two issues that become particularly relevant in this regard:

1. The use of CBRNE against people—even civilian populations—may be an element of, or a tactic in, hybrid conflict.
2. Malicious actors may attempt to distort the perceptions of another state's population in a sub-threshold conflict, for example to influence elections, or topple or destabilise governments. Spreading information to diminish trust in official authorities would be at the top of their agenda.

CBRNE has often been a part of the irregular means of conflict, as described in Chap. 1. For instance, biological warfare in terms of polluting water supplies with animal corpses has been traced back to 300 BC. Well polluting has been a widely used tactic throughout history for incapacitating populations in a geographical area. Other examples include catapulting diseased bodies or animals into walled cities to spread disease, providing smallpox blankets to Native Americans, use of chemical warfare agents during World War I and assassinations (as well as attempts) with biological, chemical, and radiological agents during and after the Cold War [27, 28, 42]. CBRNE agents have also been used against civilian populations, e.g., in Syria, and in assassinations (and attempts) such as Alexander Litvinenko, Kim Jong-nam, Sergei Skripal and Alexei Navalny in more recent times. Using CBRNE to keep a population in check, as a means of terror or to send a political signal to another state can certainly be relevant in ongoing and future conflicts (see Chap. 7).

However, the relatively "new" part of the hybrid concept may be the more complicating factor for the management of CBRNE in the future. This has been described

---

[2] Read more on how human security and global equality is—in general—constantly improving https://www.gapminder.org/.

as the convergence of physical and psychological means, kinetic and non-kinetic means, and combatants and non-combatants [16]. The main goal may not necessarily be to conquer or occupy another state, but to destabilise and influence people and societies. To decrease trust in governments and weaken the fundamentals of liberal democracies through campaigns directed at citizens may thus be a goal for threat actors. In other words, people's *perceptions* may be influenced by malicious actors and make the management of CBRNE incidents much more difficult.

Therefore, an essential focus forward should be finding ways for preventing the rapid spread of misinformation. Another scenario to consider is also preventing exploitation of a situation where malicious actors can gain momentum to turn people against the state authorities if CBRNE incidents are not managed properly.

3.1 Technological Solutions to Societal Issues?

Technology is one of the identified drivers for future development, and is predicted to further change the character and, possibly the nature of conflict [4]. Technology drives improvements, but the swift transitions also constitute a risk of social upheaval. As mentioned, technological development can aid in providing the public with the means to help themselves and others in the case of a CBRNE incident. Here, future possibilities for how technology can be used to mediate the adverse human effects of CBRNE is briefly explored. Then, some caveats and dilemmas are presented, including how the technological progress may also contribute to harm public engagement. This chapter does not, however, aim to explain the full extent of how technology can be utilised with malicious intent.[3]

Some useful tools of which we are already aware are for example video/remote diagnostics of potentially contaminated people, self-test kits for various conditions, or apps to trace the spread of chemical releases or infectious disease. Furthermore, the possibility to give people information directly to their personal or home devices provides immense opportunities. Even the increased ability to work from home due to digital development provides more opportunities for a functioning society even in states of emergency.

Some studies have suggested the potential for engaging the public in response to CBRNE incidents, meaning individual citizens as well as civic organisations. The way to engage them would be through public outreach and through communication directed at the population in question. The public would then be utilised as a capable partner in the response, as opposed to what is often perceived as a panic prone and helpless mass [10].

---

[3] For a better expansion of this topic see e.g. Cronin AK (2020) Power to the People: How Open Technological Innovation Is Arming Tomorrow's Terrorists. Oxford University Press.

Of great importance for the response in CBRNE events are, as shown in this chapter, public trust and compliance in official advice. Krieger et al. [20] describe that "...the potential of members of the public to influence the success of the overall response to CBRN releases has been acknowledged as an important factor in determining the effectiveness of the response efforts". However, they also point out that public responses have not been prioritised by scholars and practitioners, and thus the role of the public in emergency response is described as a major research and practice gap.

Nonetheless, it is quite clear that utilising civic infrastructure could relieve conventional emergency response [35, 37]. It should be possible to mobilise communities and promote self-help, such as quarantine, decontamination, or helping those at risk, in order to aid the preparedness and response effort. For this to succeed, utilising technology to better aid in building trust, creating psychological resilience and allowing people to help themselves and others is a great opportunity.

There is no definitive answer agreed upon regarding how technology affects people, or indeed how people affect technologies. It is obvious that technological solutions are only useful for preparedness purposes and emergency response if people use them and use them as intended. This means that any new technology that is meant to aid in human response must cover a perceived need (i.e., appear useful), and people must be willing and able to utilise it. Most adult individuals with medium and high living standards in the world do, indeed, own and utilise personal devices (smartphones, tablets, computers etc.) that will give them access to endless amounts of information. This could be a starting point for the public engagement.

Thus, one way to engage the public in preparedness and management is to utilise smart technology in homes and infrastructure, and to utilise personal devices and social media to reach out to people. Several, if not most, public agencies and emergency responders are utilising apps and social media to reach out to the public in general and affected groups in particular. Some examples are local police using Twitter to inform, in a given area, citizens in real time, health authorities using Instagram for public information campaigns about vaccines, or fire services using Facebook (often in combination with other means such as text messages) to reach out to inform and urge people in a certain area to stay inside and shut their windows.

Benefits of this approach is that official instances can utilise social media to inform people in a timely and effective manner when needed. Furthermore, using apps or media that allow for feedback from the public opens the possibility for frequent contact with people and two-way communication to answer questions or worries during an incident. It is also possible to trace people's movements to give them required information in case they have been in affected areas at a particular point in time. However, tracing and monitoring requires consent, safe and secure data management and strict upholding of privacy rules.

There are several ways in which social media can be useful for official bodies and emergency responders, such as the wide reach of the information to large audiences. The use of social media by officials and emergency responders also provides an opportunity for authorities to customise information and correct rumours or disinformation if needed. Using social media correctly to maintain trust and address issues and misinformation is resource demanding, so it is recommended that administering social media is a designated role in organisations that utilise it [37]. It is also necessary to ensure competent, responsible and trained administrators in order to safeguard against misuse.

It is important to remember, though, that even if many use social media as an important source for information initially, the traditional media outlets are just as important. People seek out the information sources they are used to, and most often this means the news outlets they prefer on a daily basis. The news linked to on social media often comes from "traditional" news sources as well, and people often tend to consume information from online sites, newspapers, television or radio channels after a while during an incident. For people to comply to official advice distributed via media sources, the news outlets must also be trusted.

3.2 Reliance on Technology: A Double-Edged Sword

The development of the global digital backbone whilst providing significant benefits to communities also has the potential to increase their level of vulnerability. As noted previously, the increased use of the internet and personal devices is not entirely positive [7]. It can be both an aid and a risk when it comes to informing people in a timely and correct way.

Technology can disturb and disrupt preparedness efforts or emergency response. For instance, some people may be distracted in the aftermath of an incident by looking down into the screens of their cell phones or other devices rather than paying attention to their immediate surroundings. It is also possible that there is an over-reliance on technology to communicate to the public. News blackouts can occur when cellular networks go down due to the volume of traffic, bad weather conditions, sabotage and so on.

But more importantly, when the means of communication are functioning, public advice distributed via official channels can be drowned in other, less factual, information.

There are several risks for misinformation when people choose their own sources for information. One can be misinformed by self-proclaimed "experts", who could have financial interests in promoting products that claim to counteract things such as viruses or chemical exposures. Or "armchair experts" may have developed their own theories on events and manage to present simplified, or even false, versions of reality in compelling ways to large audiences.

The spread of misinformation, or in some cases just one-sided information, can also be accelerated by media or online sites. Many of the content services record users' preferences, and then algorithms suggest similar content for them to keep consuming. This can create strong confirmation bias to the users, albeit perhaps unintentionally.

This also raises the issue of the vulnerability to the internet of things. Allowing access into our homes via devices such as Google Home, Alexa and Siri could also allow other unknown entities to pick up on preferences and leanings. And knowing such things as personal preferences allows malevolent actors to customise their information campaigns to influence our future choices, for example in elections, conflicts or purchases.

Other sources of misinformation can be false intentional data, or "fake news" that looks real to readers and recipients. There are several sources and intentions behind fake news, but in this context, it is worthwhile mentioning that this can be part of the tactics used in irregular grey zone/hybrid, conflicts. This means that an adversary is intentionally trying to mislead or influence the opinion of the whole or parts of a given population (the "target").

Conspiracy theories, fake news and other sources of misinformation are rampant on social media and online forums, and the means to combat this is of course not to impose only one source of information being available for people [6, 19]. This could perhaps be effective but would be slightly despotic. Rather, as large corporations have done lately, fact checking and addressing or deleting false information is likely a step in the right direction. Some services also offer users the possibility to receive opposite information to what they just read or watched, in order to counteract algorithms that keep feeding recipients similar information creating confirmation bias.

A further question in this regard is who checks the fact checkers? Even though "official" fact checkers have to be very transparent about their methods, they are often accused of being politicised and can be falsely discredited by trolls and conspiracists. Regardless, transparent fact checking will be required in the future and should be made as robust and incorporated into common practice as possible.

4 Organising a Responsible Preparedness

It is recommended that future research explores how "members of the public can play a constructive role in the emergency response process and promote interactions between responders and the public that treat members of the public as rationally responding actors in emergency situations" [20]. The main aim should be for the use of technology to aid in this effort. Examples could be to create tools that promote knowledge about CBRNE, apps that inform the public to create psychological resilience, customised information that automatically addresses different segments of the population or other useful features.

Utilising and involving the public could create several positive "side effects", such as increased knowledge, psychological resilience, and a more unified response from society to CBRNE incidents. Considering that communities and individuals will have access to such things as personal smart phones or smart devices at home, there are several possibilities for distributing timely and targeted risk information and crisis communication, as well as establishing a two-way connection with the civic communities. This could both be utilised to contact at risk groups in case of an emergency or engage informed volunteers to aid their neighbours or the general response effort if need be. The possibilities are many, as long as the public perceptions and needs are taken into consideration in the technology development, as well as privacy and ethical considerations.

Going back to some of the questions and ideas introduced at the start of this chapter, we can also consider who should heed this advice, and prepare for informing and engaging the public appropriately in case a CBRNE incident occurs? Even if it can be argued that international organisations, NGOs, large transnational corporations and firms etc. play a more significant role than some states in the international system [46], it is the states' responsibility to ensure their citizens' basic security and safety. Even if international regimes or organisations such as the WHO, the UN or the OPCW can aid in the management of CBRNE issues, they can hardly be held responsible for any population in any state at any time.

This chapter has underlined the importance of trust several times, and when states ignore or fail at their responsibility to handle crises (whether by brute force or softer measures), they will lose trust, further weakening any crisis management effort. When it comes to state responsibility and trust, the COVID-19 pandemic has shown how some states that have experienced one step forward in terms of welfare and governance in the recent years, experience two steps back due to poor management of a CBRNE crisis. One quote says that "The history of European state formation runs generally upward toward greater accumulation and concentration, but it runs across jagged peaks and profound valleys" [44]. The recent pandemic must be said to have constituted a profound valley for several states, both European and non-European. This shows how major the effects of a poorly managed CBRNE incident can become.

5 Conclusion

This chapter has briefly provided an overview of some important human and societal factors in CBRNE incidents, as well as some recommendations and dilemmas going forward.

How the public reacts and how public communication should be conducted from the literature showed that even if CBRNE can cause significant fear, people can still act in relatively rational ways. As long as they are provided with timely and open risk and crisis information, which is adapted to the target group, this constitutes important prerequisites for following advice. In order for the population to comply, they must trust the authorities that convey the messages. Furthermore, people should not be underestimated as assets in an emergency, and creating civic engagement can make crisis management and communication more effective.

Public communication and trust building are problematised by several factors. False information, conspiracy theories and people's lack of background knowledge are some elements that can cause adverse effects. These issues tend to be accelerated by the use of technology, as the availability of information is vast.

Furthermore, trends that indicate future development show that CBRNE elements may be truly relevant in future complex conflicts. In the case of sub-threshold, or so-called "grey zone" or "hybrid" conflicts, two challenges emerge in this context:

CBRNE may be used as a means in these types of conflicts, and there will be attempts at distorting the public's perceptions. The latter is also a significant issue in general.

However, utilising technology such as smart infrastructure, apps and social media to inform, interact with and protect people is key to succeed in future CBRNE preparedness and management. In order to help people help themselves and others, there is a unique opportunity to reach out to the public with these tools, but it must be done in a well-planned and coordinated fashion. In order to succeed with managing human and societal factors in CBRNE incidents, states must take responsibility for implementing sufficient preparedness to protect their citizens.

References

1. Barrett R, Brown PJ (2008) Stigma in the time of influenza: social and institutional responses to pandemic emergencies. J Infectious Diseases 197:34–37
2. Bellingcat (2020) Russia's clandestine chemical weapons programme and the GRU's Unit 29155. Published 23rd Oct 2020. Available via https://www.bellingcat.com/news/uk-and-europe/2020/10/23/russias-clandestine-chemical-weapons-programme-and-the-grus-unit-29155/. Accessed 22 Dec 2020
3. Cabinet Office (2011) UK Resilience: communicating risk. https://www.gov.uk/government/uploads/system/uploads/attachment_data/file/60907/communicating-risk-guidance.pdf. Accessed 9 May 2019
4. DCDC UK Ministry of Defence (2018) Global strategic trends: the future starts today. 6th ed. Available via https://assets.publishing.service.gov.uk/government/uploads/system/uploads/attachment_data/file/771309/Global_Strategic_Trends_-_The_Future_Starts_Today.pdf. Accessed 9 May 2019
5. Deer B (2003–2017) Andrew wakefield: the fraud investigation. Articles available via https://briandeer.com/mmr/lancet-summary.htm. Accessed 17 Feb 2021
6. Douglas K, Sutton RM, Cichocka A (2017) The psychology of conspiracy theories. Current Directions Psychol Sci, 538–542
7. Earnhardt R, Ackerman G (2019) Modelling terrorist technology transfer. on the horizon: security challenges at the nexus of state and non-state actors and emerging/disruptive technologies. Available via https://nsiteam.com/on-the-horizon-security-challenges-at-the-nexus-of-state-and-non-state-actors-and-emerging-disruptive-technologies/. Accessed 10 Apr 2021
8. Fischhoff B, Gonzalez RM, Small DA, Lerner JS (2003) Evaluating the success of terror risk communications. Biosecur Bioterror 1(4):255–258
9. Gigerenzer G (2006) Out of the frying pan into the fire: behavioral reactions to terrorist attacks. Risk Anal 26(2):347–351
10. Glass TA, Scoch-Spana M (2002) Bioterrorism and the people: how to vaccinate a city against panic. Confronting Biological Weapons 34:217–223
11. Hall C, Williams N, Gauntlett L et al. (2019). Findings from systematic review of public perceptions and responses. Deliverable D1.1 in PROACTIVE project. Available via https://proactive-h2020.eu/wp-content/uploads/2020/01/proactive_20191223_D1.1_V4_PHE_Systematic-Review-of-Public-Perceptions-and-Responses.pdf. Accessed 10 Apr 2021
12. Hall RCW, Chapman MJ (2006) Medical and psychiatric casualties caused by conventional and radiological (dirty) bombs. Gen Hospital Psychiatry 28:242–248
13. Harrison AA, Moulton TJ (1997) The Kennedy assassination, unidentified flying objects, and other conspiracies: psychological and organizational factors in the perception of "cover-up.' Systematic Res Behav Sci 14(2):113–128

14. Heath RL (2006) Best practices in crisis communication: evolution of practice through research. J Appl Commun Res 34(3):245–248
15. Henderson JN, Henderson L, Carson RGE, Boatright DT (2004) Chemical (VX) terrorist threat: public knowledge, attitudes, and responses. Biosecur Bioterror 2(3):224–228
16. Hoffman FG (2009) Hybrid warfare and challenges. Joint Force Quart 52:34–39
17. IAEA (International Atomic Energy Agency) (1988) The radiological accident in Goiânia. https://www-pub.iaea.org/MTCD/publications/PDF/Pub815_web.pdf. Accessed 11 Aug 2020
18. Jolley D, Douglas KM (2014) The effects of anti-vaccine conspiracy theories on vaccination intentions. PLOS ONE 9(2):e89177. Available via https://doi.org/10.1371/journal.pone.0089177. Accessed 2 Dec 2020
19. Jolley D, Douglas KM (2014) The social consequences of conspiracism: Exposure to conspiracy theories decreases the intention to engage in politics and to reduce one's carbon footprint. Br J Psychol 105:35–56
20. Krieger K, Amlôt R, Rogers MB (2014) Understanding public responses to chemical, biological, radiological and nuclear incidents—driving factors, emerging themes and research gaps. Environ Int 72:66–74
21. Lindell MK, Perry RW (2012) The protective action decision model: theoretical modifications and additional evidence. Risk Anal 32(4):616–632
22. Liu D, Asgharpour F, Camp LJ (2008) Risk communication in security using mental models. In: Usable Security. http://usablesecurity.org/papers/liu.pdf. Accessed 4 Jul 2019
23. McCauley M, Minsky S, Viswanath K (2013) The H1N1 pandemic: media frames, stigmatization and coping. BMC Public Health 13:1116. Available via http://www.biomedcentral.com/1471-2458/13/111. Accessed 2 Dec 2020
24. McComas KA (2006) Defining moments in risk communication research: 1996–2005. J Health Commun Int Perspect 11(1):75–91
25. National Intelligence Council (NIC) (2017) Global trends: paradox of progress. Available via https://www.dni.gov/files/documents/nic/GT-Full-Report.pdf. Accessed 9 May 2019
26. NATO Allied Command Transformation (ACT) (2017) Strategic foresight analysis 2017 report. https://www.act.nato.int/images/stories/media/doclibrary/171004_sfa_2017_report_hr.pdf. Accessed 9 May 2019
27. Parachini J (2001) Anthrax attacks, biological terrorism and preventive responses. Testimony of John parachini, policy analyst, RAND Washington Office, before the subcommittee on technology, terrorism, and government information, Nov 6, 2001. RAND
28. Poupard JA, Miller LA (1992) History of biological warfare: catapults to capsomeres. Ann NY Acad Sci 666:9–20
29. Quarantelli EL (2001) The sociology of panic. Disaster Research Center, University of Delaware. http://dspace.udel.edu/handle/19716/308. Accessed 9 Jan 2019
30. Quinn P (2018) Crisis communication in public health emergencies: the limits of 'Legal Control' and the risks for harmful outcomes in a digital age. Life Sci Soc Policy 14(4). Available via https://pubmed.ncbi.nlm.nih.gov/29404722/. Accessed 9 Jun 2019
31. Reidpath DD, Chan KY, Gifford SM, Allotey P (2005) 'He hath the French pox': stigma, social value and social exclusion. Sociol Health Illn 27(4):468–489
32. Rogers MB, Amlôt R, Rubin GJ (2013) The impact of communication materials on public responses to a radiological dispersal device (RDD) attack. Biosecur Bioterror 11(1):49–58
33. Rogers MB, Amlôt R, Rubin GJ, Wessely S, Krieger K (2007) Mediating the social and psychological impacts of terrorist attacks: the role of risk perception and risk communication. Int Rev Psychiatry 19(3):279–288
34. Rubin GJ, Chowdhury AK, Amlôt R (2012) How to Communicate with the public about chemical, biological, radiological or nuclear terrorism: a systematic review of the literature. Biosecur Bioterror 10(4):383–395
35. Scoch-Spana M, Franco C, Nuzzo JB, Usenza C (2007) Leadership tool for catastrophic health events. Biosecur Bioterror 5(1):8–25
36. Sheppard B, Rubin GJ, Wardman JK, Wessely S (2006) Terrorism and dispelling the myth of a panic prone public. J Pub Health Policy 27:219–245

37. Simona T, Goldberg A, Adini B (2015) Socializing in emergencies—a review of the use of social media in emergency situations. Int J Inf Manage 35:609–619
38. Sorensen JH (2000) Hazard warning systems: review of 20 years of progress. Nat Hazard Rev 1(2):119–125
39. Spencer ML, Kindt MT, Stans MP (2011) Public Resilience in CBRN events: lessons learned from seven cases. The counterproliferation papers future warfare series 52. USAF Counterproliferation Center
40. Stern J (2002) Dreaded risks and the control of biological weapons. Int Secur 27(3):89–123
41. Swain KA (2012) Explanation of risk and uncertainty in news coverage of an anthrax attack. J Risk Anal Crisis Response 2(2):81–95
42. Szinicz L (2005) History of chemical and biological warfare agents. Toxicology 214:167–181
43. Tanaka Y (1998) Psychological dimensions of risk assessment: risk perception and risk communication. Prog Nucl Energy 32(314):243–253
44. Tilly C (1992) Cities and states in world history. In: Tilly C (ed) Coercion, Capital and European State, AD 990–1992. Blackwell, Oxford, pp 1–37
45. Wessely S (2005) Don't panic! Short- and long-term psychological reactions to the new terrorism: the role of information and the authorities. J Mental Health 14(1):1–6
46. Willetts P (2008) Transnational actors and international organizations in global politics. In: Baylis J, Smith S, Owens P (eds) The globalization of world politics, 4th edn. Oxford University Press, p 33

[illegible]

Risk in the Infrastructure of the Future: A Holistic View

N. Goldbeck and W. Y. Ochieng

1 Introduction

The man-made systems that our societies fundamentally depend upon are commonly referred to as critical infrastructure. They include transportation networks, energy systems, communication services, water supply, waste disposal, flood protection, health and social care, emergency and security services, financial institutions and supply chains for essential goods. Critical infrastructure systems require vast physical assets that can be described as networks with large geographic coverage and highly connected hub locations. Organisations that own and operate critical infrastructure are often under direct government control or strictly regulated. In the UK, it is estimated that 10 million people or about one third of the total workforce are employed in critical infrastructure sectors [1].

To assist how societies can better protect their infrastructure against the changing nature of CBRNE threats and the geopolitical developments, it is vital that we understand the evolving infrastructure systems in terms of their increasingly cyber-physical interconnectedness and the benefits and vulnerabilities they present. Conventional protection strategies are at risk of becoming inefficient or even ineffective, due to several shortcomings highlighted in this chapter. We also discuss three emerging concepts that contribute to a more holistic view: resilient infrastructure systems, systematic interdependency analysis and cyber-physical security. The last part of the chapter will outline a novel risk assessment framework to meet the future challenge.

N. Goldbeck · W. Y. Ochieng (✉)
Department of Civil and Environmental Engineering, Imperial College, London SW7 2AZ, UK
e-mail: w.ochieng@imperial.ac.uk

N. Goldbeck
e-mail: n.goldbeck14@imperial.ac.uk

P.D.E. Biggins and D. Chana (eds.), *CBRNE: Challenges in the 21st Century*, Advanced Sciences and Technologies for Security Applications,
https://doi.org/10.1007/978-3-031-17374-5_9

2 Siloed Risk Management

In many ways, conventional infrastructure protection strategies are biased towards a siloed risk management approach. Current thinking on critical infrastructure protection (CIP) is that it can be viewed as a risk management process, and as such depends on how we conceptualise risk. Risk can be defined as a product of three factors:

$$\text{Risk} = \text{Hazard} \times \text{Exposure} \times \text{Vulnerability}$$

Another widely used definition focusses on expected outcomes, rather than sources of risk:

$$\text{Risk} = \text{Consequence} \times \text{Probability}$$

Most standardised risk management frameworks feature a risk identification step, in which the aim is to find all relevant risks. For example, the international standard on risk management ISO 31000 [2], used by many organisations responsible for infrastructure assets and operations to develop their risk management processes, features risk identification as the first step of the risk assessment which then informs the selection of appropriate risk treatment options (Fig. 1). The output of the risk assessment is a risk register and is often visualised as a risk matrix, a two-dimensional plot of the impact and likelihood of all items in the risk register.

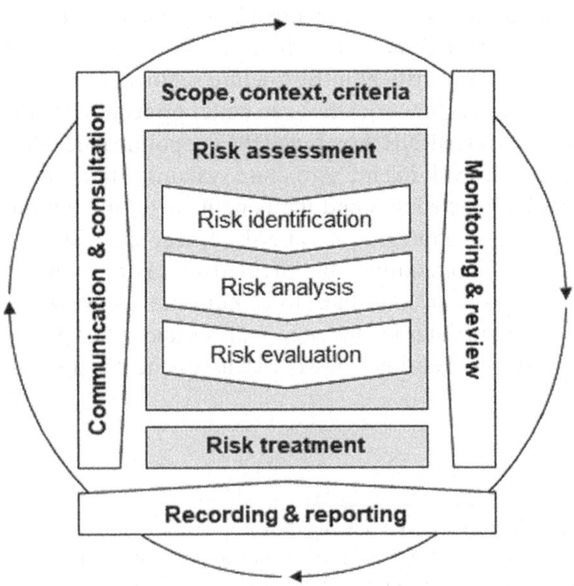

Fig. 1 Illustration of the ISO 31000 risk management framework

Like most state-of-the-art frameworks, the ISO 31000 standard assumes that the space of overall risk can easily be discretised,[1] and it suggests analysing and treating risks individually. In other words, the conventional approach to risk management is a divide-and-conquer strategy. It has proven highly effective in controlling known risks and its implementation is relatively straight forward, even in large organisations facing many different risks. However, the divide-and-conquer strategy also has some disadvantages that are worth discussing, especially as they become more and more relevant with increasing complexity of the risk management problem.

In practice, one often finds that it is relatively easy to identify the main types of risk and to describe them at a high level. But when it comes to concrete impact scenarios, there are usually too many variations of how hazardous events could play out and it is not possible to include all combinations of root causes and contributing factors in the risk register. Many emergency plans could be frustrated simply by the concurrency of two hazards that had been considered separately in the risk management, for example, a physical attack with an explosive device and a cyber-attack that prevents a coordinated response. Another issue is the possibility of higher order effects or cascading failure which makes it difficult to estimate the overall consequences of a disruption and prioritise mitigation measures adequately.

In mathematical terms, this means that critical infrastructure risk is an extremely high-dimensional space and that a comprehensive risk model would have to contain a very large number of variables to capture all possible variations of hazards as well as the dynamic state of the infrastructure systems that determines their exposure and vulnerability. Standard risk management approaches (explicitly or implicitly) use dimensionality reduction techniques, discretisation and sampling of the risk space in order to obtain a finite and manageable set of individual risks to be included in the risk register. This means that discretisation and sampling errors have to be considered, especially if there are non-linear effects that determine the system's response.

Aggregating risk factors into categories and discretising continuous variables is impossible to avoid due to the complexity and uncertainty of the risk management problem. But we should be aware that this process always comes at the cost of losing some information. In practice, the degree of this information loss is usually determined by prior decisions on resource allocation for the risk management rather than specific objectives regarding the coverage or confidence of the risk assessment. The methods used in this important step of critical infrastructure protection are often not sufficiently verified and validated.

There is a danger that organisations are led to believe that by following the guidelines of a standardised risk management framework they will obtain a *complete* record of the entire risk space. And even if organisations are aware that the risk register is the result of a selection process and only a more or less representative sub-set of risks, there is usually very little visibility of the domain-specific knowledge and assumptions that this selection process is based on. To avoid a false sense

[1] Discretisation is "the process of transferring continuous functions, models, variables, and equations into discrete counterparts. This process is usually carried out as a first step toward making them suitable for numerical evaluation and implementation on computers" [3].

of security, organisations should consider the updating of their risk register a continuous process and actively seek to identify gaps, potentially with the help of external experts. Regular knowledge sharing and coordination between different business and government departments will also help to avoid the pitfalls of compartmentalised risk intelligence.

Another problem is that conventional risk management methods are biased towards assuming that the identified risks are independent of each other, although there may exist significant dependencies or even interdependencies between them. The probability of different hazard events can be correlated positively or negatively if there are causal links or common dependencies on certain states of the system or its environment. For example, a large public event always increases the risk in multiple dimensions, such as posing a more attractive target and making real-time monitoring more difficult.

Further interdependencies can exist within the infrastructure system itself between different assets and/or operational processes. For instance, digital building information models (BIM) can help to coordinate the response to a security incident. However, if such digital assets become inaccessible due to a network failure, power failure or a targeted cyber-attack the incident response may be significantly delayed.

Interdependencies can also exist between the treatment options chosen for different risks. For example, access control to certain areas with extensive security screening induces the risk of large queues and the entry points which itself may become targets for malicious attacks. Ignoring such interdependency effects can render risk management strategies inefficient or even ineffective in the worst case.

Finally, state-of-the-art risk management frameworks are tailored to an individual organisation seeking to control the risk to their assets and operations. However, in the critical infrastructure sector the responsibility for CBRNE risk management is usually shared between different entities, including private companies and government agencies. In the UK, the Centre for the Protection of National Infrastructure (CPNI) and the National Cyber Security Centre (NCSC) provide advice and support to stakeholders in the various infrastructure sectors. Such technical authorities are important facilitators for knowledge sharing between decision makers in the private and public sectors. However, the extent to which infrastructure companies are able and incentivised to invest in protective security is directly dependent on the regulatory environment and there remain open questions on how the cost of security should be shared to achieve the common goal of reducing overall societal risk.

To some extent, the shortcomings of current risk management approaches in the domain of critical infrastructure protection can be mitigated by adapting existing processes. It helps to think of the risk identification step as a process to maximise coverage over the high-dimensional space of risk and to keep records of the domain knowledge and assumptions that inform this process. Discretisation errors can be minimised in some cases by conducting a sensitivity analysis and using the results to select an appropriate level of granularity. Hierarchical approaches, such as the distinction between operational, tactical and strategic risks, can help to divide responsibilities and clearly communicate priorities when the scope of a risk management problem in the critical infrastructure domain is very large. However, such hierarchies also introduce additional interfaces between the different levels and require good planning and co-ordination.

In the following sections, we will discuss some emerging concepts that promote a more holistic approach to critical infrastructure protection. However, it is not suggested that these should replace established practices of risk management, but rather that they are to be seen as additional methods that help to better address new challenges around complexity and interconnectedness. The fact that these concepts have gained much attention in recent years from academic researchers and industry practitioners alike highlights the need for further developing the toolbox for critical infrastructure protection.

3 How the Concept of Resilience Facilitates More Holistic Approaches

Infrastructure resilience is an emerging concept that aims to strengthen the capability of infrastructure systems to cope with disruptions. This addresses the problem of gaps in conventional risk management strategies related to high-impact, low-probability risks that cannot be completely avoided, controlled or transferred (at least within practical budget constraints), and sometimes not even anticipated.

Analysing the resilience properties of infrastructure systems has been influenced by studies in the field of ecology where resilience is defined as follows:

> Resilience determines the persistence of relationships within a system and is a measure of the ability of these systems to absorb changes of state variables, driving variables, and parameters, and still persist.
>
> Holling [4]

Critical infrastructure systems are considered resilient if they are able to minimise the impact of disruptions and quickly recover the residual loss of system performance. Thus, infrastructure resilience can be defined as:

> The joint ability of infrastructure systems to resist (prevent and withstand) any possible hazards, absorb the initial damage, and recover to normal operation.
>
> Ouyang et al. [5]

The concept of resilience has also been extended to entire cities and is then referred to as urban resilience, meaning:

> The capacity of individuals, communities, institutions, businesses, and systems within a city to survive, adapt, and grow no matter what kinds of chronic stresses and acute shocks they experience.
>
> 100 Resilient Cities [6]

As the functioning of a city critically depends on the state of its infrastructure systems, *urban* resilience includes *infrastructure* resilience.

There is considerable overlap between the definitions of resilience and reliability, with the latter being defined as "the ability of an item to perform a required function, under given environmental and operational conditions and for a stated period of time" [7]. However, several important resilience issues are not addressed by classical reliability engineering, such as decisions taken during the recovery period or trade-offs between robustness and recovery capabilities. Thus, in the same way that the resilience concept extends conventional risk management approaches it also extends classical reliability theory.

The specific capabilities identified by qualitative research as resilience strengthening broadly fall into four categories:

- Robustness capabilities that help systems to withstand or absorb perturbations.
- Redundancy capabilities that allow systems to maintain service levels though the activation of reserve capacity.
- Recovery capabilities that enable systems to repair or replace damaged assets quickly.
- Adaptation capabilities that help systems to operate in changed environmental conditions.

Qualitative resilience concepts provide guidance on how to design systems and plan processes to become better in these resilience dimensions. A common decision problem is how limited resources should be allocated to different resilience measures and how to take into account trade-offs, for example, between strengthening robustness and recovery capabilities. For this class of problems, it can be helpful to quantify the resilience of different systems or system configurations.

An example for a resilience quantification method is the City Resilience Index (Fig. 2) developed by Arup [8]. This multi-level assessment framework consists of four dimensions, 12 goals, 52 indicators and 156 variables. Such resilience assessment frameworks systematically aggregate knowledge about what makes a particular type of system resilient. They are particularly useful for benchmarking studies and what-if analysis. However, it is important to keep in mind that composite indices are based on prior knowledge and subjective decisions regarding the selection and weighting of performance areas and indicators. Therefore, they are unlikely to lead to the discovery of previously unknown resilience phenomena.

Another approach to quantifying resilience is the use of a system performance measure. This reflects the fact that many resilience definitions are formulated around the system's ability to maintain and recover its core functionality. We assume that the function $P(t)$ measures the performance of an infrastructure system and is normalised to values between 0 and 1. For a specific hazard scenario, the resilience loss metric [9] is then defined as:

$$\mathrm{RL} = \int_{t_0}^{t_\mathrm{R}} (1 - P(t))\mathrm{d}t$$

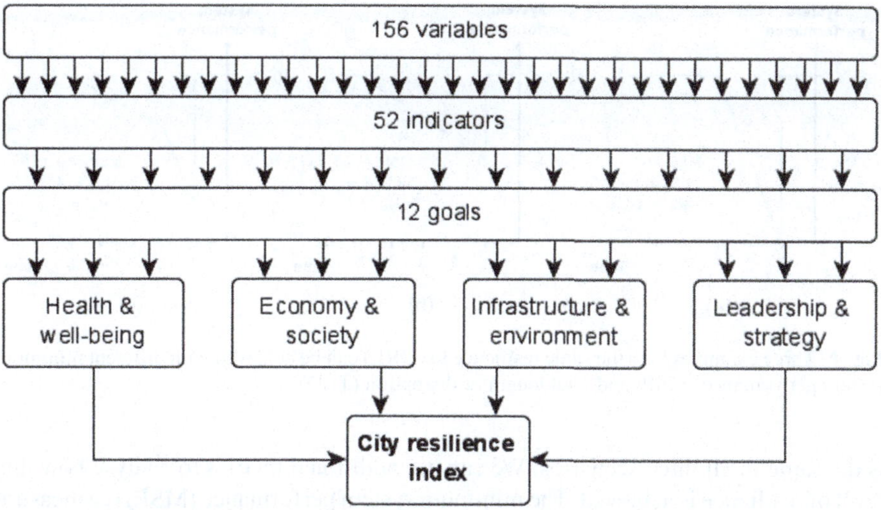

Fig. 2 The city resilience index [8] as an example of a composite resilience metric

where t_0 is the time when the disruptive event begins and t_R is the time when the system fully recovered its pre-disruption performance level. In an idealised case where the performance loss is a step function and recovery a linear process, the resilience loss metric corresponds to the area of the so-called resilience loss triangle (Fig. 3).

A system performance function also allows to assess two key components of resilience, namely robustness and recovery capability, separately. Figure 4 illustrates three different scenarios how a system could react to a disruption. The resilience loss

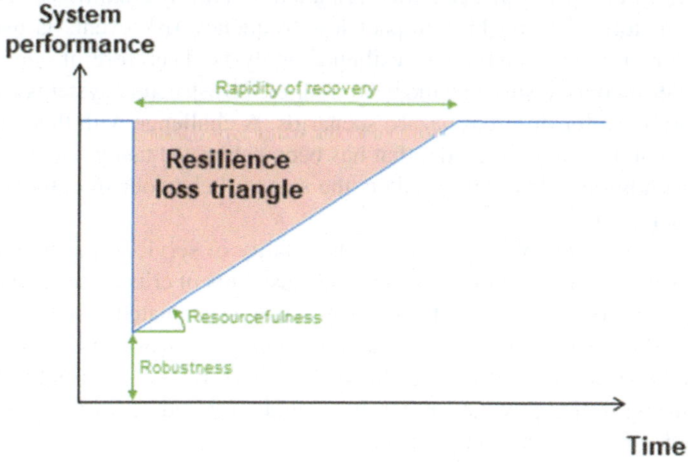

Fig. 3 Measuring resilience with a system performance function

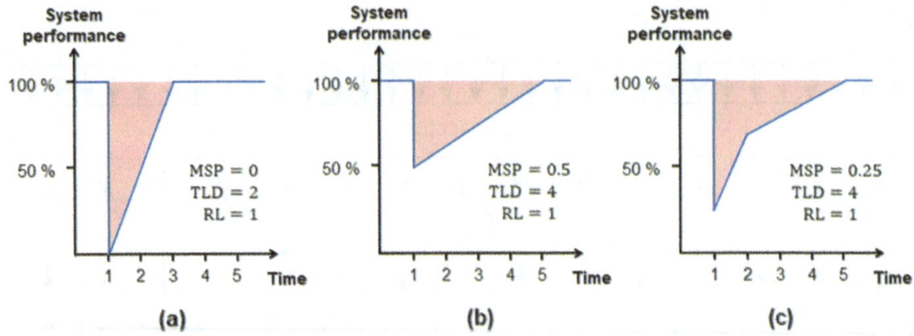

Fig. 4 Three examples how the same resilience loss (RL) can be achieved with different minimum system performance (MSP) and total length of disruption (TLD)

is the same in all three scenarios. We can use additional metrics to analyse how this level of resilience is achieved. The minimum system performance (MSP) is a measure for the robustness of the system. The total length of disruption (TLD) is a measure for the recovery capability of the system. In the example, system (a) achieves resilience by virtue of a high recovery capability. It is twice as fast to fully recover than the other two systems. However, it has the lowest robustness and becomes completely inoperable at the beginning of the incident. In contrast, system (b) achieves resilience by means of a high robustness. Its performance never drops below 50%. The recovery of system (c) is very fast initially, but then slowing down such that the TLD is the same as for system (b).

An important advantage of performance-based resilience metrics is that they do not depend on prior knowledge regarding the importance of certain resilience mechanisms. However, they do require a function that sufficiently captures the performance of the entire system. For empirical studies, it is important to have sufficient data to evaluate the system performance under a range of operating conditions. As per their definition, historic data on high-impact low-frequency risk events is usually not sufficient for in-depth quantitative resilience analysis. Therefore, it is common to use computer models to simulate the behaviour of infrastructure systems and predict their performance for different hazard scenarios. A challenge with this approach is to validate that a simulation model that has been calibrated using data from normal operating conditions accurately predicts the system behaviour in entirely different operating conditions.

In most cases, available capacity and the quality of services are suitable performance measures for quantitative resilience assessments of critical infrastructure. For example, the performance of an electricity grid can be evaluated based on how much of the overall electricity demand it is able to supply to customers, as measured in the form of real or simulated power flows. The performance of a public transport network during a disruptive event can be evaluated based on excess journey time caused by delayed or cancelled journeys.

Adopting resilience approaches for protecting critical infrastructure systems from CBRNE threats means developing a detailed understanding of how the systems would react under potentially challenging hazard situations. Dependent on whether it is a C, B or R event or a combination, different approaches will need to be adopted to handle the hazard situation. The focus on identifying system capabilities required during different phases of a disruptive incident encourages a holistic view when it comes to developing strategies to improve the resilience of critical infrastructure.

An important part of most resilience strategies are measures to improve collaboration and information sharing between different stakeholders. To this end, it is essential to have a common language to describe CBRNE hazards and their impacts. Information sharing is important before, during and after a disruptive event and can take many forms, for example, briefings, training events, and sharing of documents, data or application programming interfaces (APIs). It can be useful to organise information sharing around best practices for mitigating particular types of threats. However, resilience practitioners often find that the greatest learning effects materialise around general issues, for example, how responsibilities for continuous organisational development should be assigned. It is important to establish forums and communication channels that allow different stakeholders to communicate relevant information freely while avoiding the risk of sensitive data being obtained by unauthorized parties. When infrastructure companies from the private sector are involved, additional complexity is added by the fact that sharing certain data could result in loosing competitive advantages and may be conflicting with their business interests.

Another aspect of how resilience approaches encourage a more holistic view is the identification of trade-offs. Consider, for example, the risk of a power outage at an airport. A conventional risk management approach would seek to achieve a level of reliability that minimises the power outage risk, for example, with an additional, independent connection to the grid. Resilience thinking means to consider allowing for some level of disruption if that helps to guarantee the persistence of the system's core functionality. For instance, another mitigation option for the airport might be an on-site emergency power system for safety-relevant equipment and a contractual arrangement with a third party that can bring additional emergency generation equipment for non-critical equipment on site within a certain response time. A comprehensive resilience analysis can identify all such mitigation options and help to select the ones that will minimise the expected overall cost of the disruption.

As we have already seen above, there are many ways in which infrastructure systems can be resilient. Learning about best practices in other industries and organisations is often beneficial but it is not always guaranteed that a solution made for one system will bring more resilience to another. Therefore, all critical infrastructure systems should undergo regular stress testing in simulation to validate that they meet minimum resilience standards. A basic workflow for stress testing critical infrastructure is depicted in Fig. 5. To better manage the complexities of large infrastructure systems, multi-level stress testing methodologies have been developed that combine system- and component-level performance assessments [10].

In the UK, the National Infrastructure Commission recommends that the government should facilitate and regulate the stress testing of critical infrastructure [11].

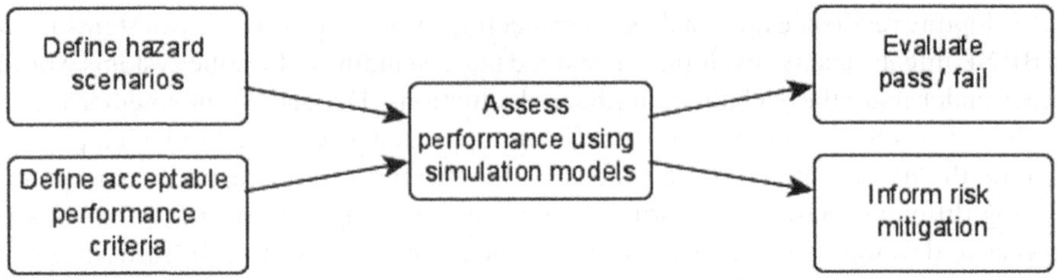

Fig. 5 Stress testing workflow

Infrastructure systems have been resilient to several shocks and stresses over recent years but a more proactive approach is needed to maintain this level of resilience, considering the possibilities of different and harder challenges in the future. Currently, each individual operator decides independently whether to conduct stress testing or not, which scenarios to test against and how to evaluate resilience performance. The National Infrastructure Commission recommends that regulators should require energy, water, digital, road and rail infrastructure operators to do regular stress testing by 2024, provide guidance on the scope and scrutinise the outcomes. However, carrying out the stress testing should be left to the operators, as they have the best understanding of their systems.

In the finance industry, it was only after the crisis of 2008 that regulators started developing common frameworks for stress testing banks. It is to hope the critical infrastructure sectors will adopt stress testing before a major disaster occurs and uncovers gaps in critical infrastructure protection strategies that could have been identified with a comprehensive resilience analysis.

Of course, improved resilience also comes at a cost. The National Infrastructure Commission recognises that resilience must be appropriately valued in regulatory price reviews and cannot be left entirely to the market. The development of resilience standards should be informed by a comprehensive assessment of the costs and benefits, which are often not limited to just an individual infrastructure operator and its customers. Moreover, resilience standards should be based on realistic assumptions on how the government could help respond to catastrophic infrastructure failure and how the public would be able to cope with major service disruptions in different sectors.

4 How a Systematic Analysis of Interdependencies Addresses the Complexity Challenge

Infrastructure interdependency can be defined as "a bidirectional relationship between two infrastructures through which the state of each infrastructure influences or is correlated to the state of the other" [12]. Empirical evidence for such interdependencies has been found in many cases of infrastructure failure. For example, a study of the 2001 World Trade Center attack estimated that 20% of the resulting infrastructure disruptions across eight different systems were caused by interdependency effects, with co-location of assets being the most frequent cause [13]. Interdependency can also be observed during the recovery process. For example, the rate at which services were restored in communication, power and water networks after the 2010 earthquake in Chile exhibited significant cross-correlations [14]. In particular, it was found that a quick restoration of communication services facilitates the recovery in other infrastructure systems but in turn depends on resilient power supply.

The method of systematically identifying all relevant dependency relations between a set of infrastructure systems is referred to as interdependency mapping. Figure 6 depicts an example of an interdependency map for critical infrastructure systems in a metropolitan region. A simple topological analysis on this graph, for example, ranking the nodes by the number of incoming and outgoing edges (degree centrality) can give an indication of how important each system is for the overall functioning of the system-of-systems.

There are different types of dependency relations between infrastructure systems. A taxonomy like the one presented in Table 1 can guide the interdependency mapping process. However, there are no standard definitions for dependency types, and some are known under different names in the literature. Therefore, it is recommended to begin the systematic interdependency analysis with clearly defining the dependency types as suitable for the particular use case.

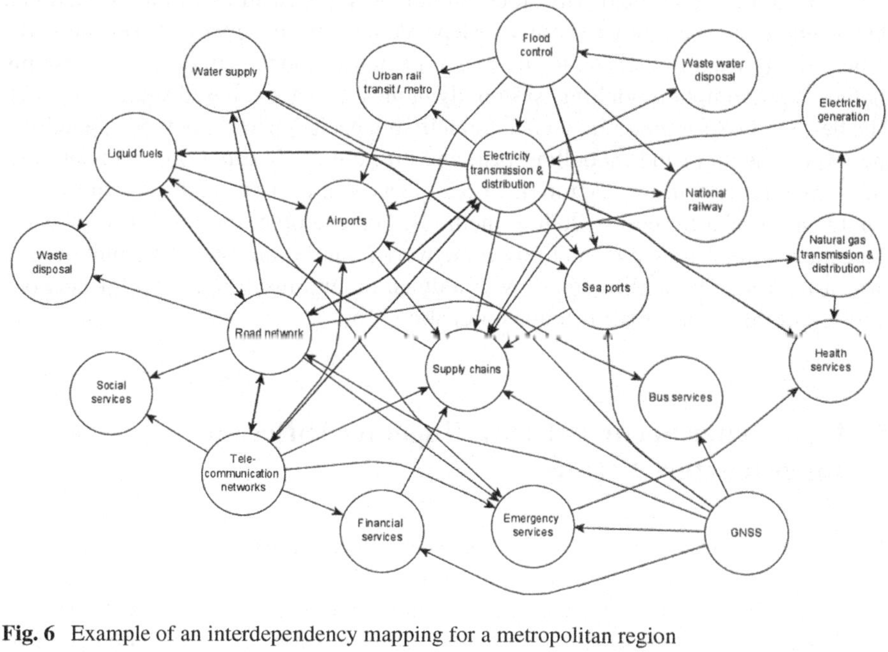

Fig. 6 Example of an interdependency mapping for a metropolitan region

Table 1 Common types of dependencies between critical infrastructure systems

Dependency type	Description	Example
Input, Physical, Functional	A system uses (material) outputs of another system	A metro system requires electricity supply
Geographic, Spatial, Co-location	Multiple systems are situated in the same environment and affected by local events	District heating pipelines and telecommunication cables are running in parallel
Cyber, Informational	A system is dependent on information transmitted through another system	An electricity network depends on a SCADA system
Policy, Procedural	A system is dependent on policies and procedures in another system	A transport service is disrupted because of a police operation
Societal	A system is dependent on societal factors that can change due to events in other systems	Travel demand changes because of public fear after a terrorist attack
Recovery	A system's recovery depends on the state of another system	Emergency engineering staff cannot access damaged assets due to road obstructions

Identifying and mapping dependency relations is also a prerequisite for creating a simulation model that can be used to analyse the resilience of a system-of-systems. The geographic correlation of component failure in multiple networks due to co-location can be represented with stochastic models of cascading failure. Functional dependencies can be captured with interdependency links in network-based methods. Other widely used modelling techniques for interdependent infrastructure systems include input–output modelling, system dynamics, and agent based simulation [15]. The best method for modelling and parametrising each dependency type depends on the requirements of the modelling task and on how the internal processes of each infrastructure system are captured. In many cases, a holistic analysis requires the combination of different modelling methods. For example, the resilience of interdependent public transport and power networks can be analysed by combining a stochastic asset operability model with a dynamic optimisation model for network flow assignment and repair resource allocation [16].

5 How Cyber-Security Draws Together Different Infrastructure Sectors

Across all infrastructure systems, the digital transformation has been the most significant technological development in recent times. The trend towards more and more connected devices that exchange information via the internet is remarkably similar in all sectors. Examples include the smart phone, smart grid, smart home, smart mobility and so on. There are already more devices connected to the internet-of-things (IoT) than people on the planet. By 2030 the number of connected devices is expected to exceed 50 billion.

Considering the enormous damage that critical infrastructure outages could have, it comes as no surprise that these systems have already been the target of many cyber-attacks. Security experts are particularly concerned about hackers infiltrating industrial control systems. The real possibility of such a scenario has been demonstrated, for example, in 2015, when hackers caused a system-wide power blackout lasting several hours in the Ukraine by accessing the supervisory control and data acquisition system (SCADA) and triggering circuit breakers. Major disruptions have also been caused by ransomware attacks, for example, on the National Health Service in England or the Colonial Pipeline for refined oil products in the US. Another concern is that cyber-attacks could be used to facilitate physical attacks by means of exposing security-relevant information (as was the case when computers in the Indian nuclear power plant Kundankulam were infected with malware) or manipulating existing control measures (e.g. CCTV, access control).

Many national governments have established cyber security centres to advise stakeholders in the public and private sectors on how to improve their resilience against cyber-attacks. For example, the UK's National Cyber Security Centre provides a Cyber Assessment Framework for critical infrastructure operators which defines 14 principles that can be assessed using a collection of indicators of good practice [17].

Cyber-security is certainly one of the biggest challenges for critical infrastructure protection and governments and private-sector operators must scale up their investment in skilled staff, better software and third-party services in order to keep up with the ever more sophisticated technological capabilities of adversaries. The convergence of different critical infrastructure systems towards more and more common technologies (e.g. internet protocols, encryption standards) and sometimes even shared computing infrastructure increases the stakes but also means that there are more and more potential synergies of which to be taken advantage.

6 An Integrated Risk and Resilience Framework for Critical Infrastructure Protection

This chapter has argued that conventional methods are no longer sufficient to protect critical infrastructure from CBRNE threats due to fundamental changes regarding the potential hazards, geopolitical climate and infrastructure systems themselves. We have discussed emerging, more holistic approaches, like infrastructure resilience, system-of-systems analysis and cyber security strategies. At the same time, it is important not to throw overboard the proven and tested methods of conventional risk management. But how can all these approaches work together?

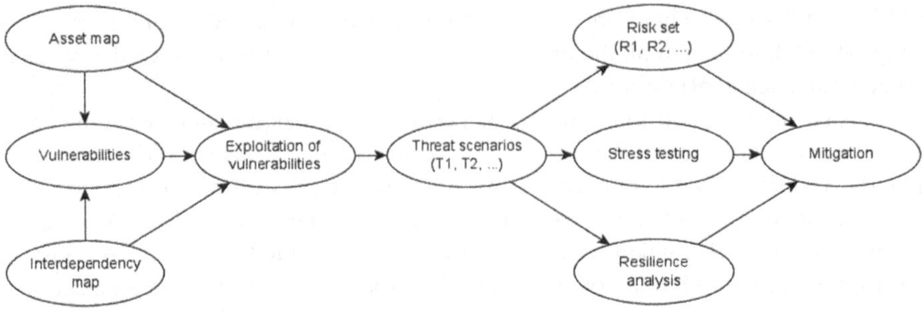

Fig. 7 Proposed framework for integrated risks management, resilience analysis and stress testing

In Fig. 7, we outline a general framework for integrating the key approaches discussed in this chapter, namely risk management, resilience analysis, stress testing and interdependency analysis. There are some fundamental concepts that remain important no matter which combination of tools is used to address the challenge of future CBRNE threats. The first is that critical infrastructure protection can only succeed if it is based on a solid understanding of how the infrastructure systems function and how they interact with each other and their environment. Therefore, the starting point of the proposed framework is the comprehensive mapping of all infrastructure assets. Crucially, this should include physical, institutional and digital assets. In a second step, a database of dependency relations between these assets is created. Together, these databases enable an accurate description of the vulnerabilities in the system-of-systems.

Vulnerabilities translate into threats if they can be exploited. A detailed description of threat scenarios is an input for the three subsequent steps: risk assessment, stress testing, and resilience analysis. These three steps could be carried out separately. However, in practice we often find that there are synergies between them because they all require models to predict the performance of infrastructure after disruptive events. Finally, the design of a mitigation strategy involves selecting a suitable combination of risk treatment and resilience strengthening measures. This is usually done based on a comprehensive cost benefit analysis, where the benefits comprise the risk-reduction effects estimated based on the preceding analysis.

7 Conclusion

There are many emerging issues around the changing nature of CBRNE threats for cyber-physical systems in a world where new technologies are disseminated faster than ever before, information and data can be made available anywhere around the globe within milliseconds, and our safety and wellbeing depend on infrastructure assets that are all connected to one unified global communication system. Manuel Castells predicted in his ground-breaking book *The Rise of the Network Society* that ubiquitous information and telecommunication technologies would fundamentally change how our social structures and activities are organised [18]. Through academic research, but unfortunately also through experience at the cost of lives, the world began to understand that established security concepts are not sufficient to ensure everyone's safety in the new network society. Networks can be the most fragile structures with single points of failure and the risk of cascading effects leading to catastrophic outcomes. Fortunately, networks can also be the most resilient structures, self-healing and with numerous redundant connections. It all depends on how we design them.

References

1. UK Office for National Statistics (2020) Key workers reference tables. https://www.ons.gov.uk/employmentandlabourmarket/peopleinwork/earningsandworkinghours/datasets/keyworkersreferencetables. Accessed 15 Dec 2020
2. ISO (2018) ISO 31000:2018: Risk management—guidelines. https://www.iso.org/standard/65694.html. Accessed 12 Nov 2020
3. Wikipedia Discretization. https://en.wikipedia.org/wiki/Discretization. Accessed 15 Dec 2020
4. Holling CS (1973) Resilience and stability of ecological systems. Annu Rev Ecol Syst 4:1–23
5. Ouyang M, Dueñas-Osorio L, Min X (2012) A three-stage resilience analysis framework for urban infrastructure systems. Struct Saf 36–37:23–31. https://doi.org/10.1016/j.strusafe.2011.12.004
6. 100 Resilient Cities (2016) What is urban resilience? http://www.100resilientcities.org/resilience. Accessed 24 Mar 2019
7. Rausand M, Høyland A (2004) System reliability theory: models, statistical methods, and applications, 2nd edn. Wiley
8. Arup (2014) City resilience index: city resilience framework. https://www.arup.com/perspectives/publications/research/section/city-resilience-index. Accessed 22 Mar 2019
9. Bruneau M, Chang SE, Eguchi RT, Lee GC, O'Rourke TD, Reinhorn AM, Shinozuka M, Tierney K, Wallace WA, von Winterfeldt D (2003) A Framework to quantitatively assess and enhance the seismic resilience of communities. Earthq Spectra 19:733–752. https://doi.org/10.1193/1.1623497
10. Esposito S, Stojadinović B, Babič A, Dolšek M, Iqbal S, Selva J, Broccardo M, Mignan A, Giardini D (2020) Risk-based multilevel methodology to stress test critical infrastructure systems. J Infrastr Syste 26. https://doi.org/10.1061/(ASCE)IS.1943-555X.0000520
11. UK National Infrastructure Commission (2020) Anticipate, react, recover: resilient infrastructure systems. https://nic.org.uk/studies-reports/resilience/. Accessed 15 Dec 2020
12. Rinaldi SM, Peerenboom JP, Kelly TK (2001) Identifying, understanding, and analyzing critical infrastructure interdependencies. IEEE Control Syst Mag 21:11–25
13. Mendonça D, Wallace WA (2006) Impacts of the 2001 world trade center attack on New York city critical infrastructures. J Infrastruct Syst 12:260–270
14. Dueñas-Osorio L, Kwasinski A (2012) Quantification of lifeline system interdependencies after the 27 February 2010 Mw 8.8 Offshore Maule, Chile. Earthquake. Earthq Spectra 28:S581–S603. https://doi.org/10.1193/1.4000054
15. Ouyang M (2014) Review on modeling and simulation of interdependent critical infrastructure systems. Reliab Eng Syst Saf 121:43–60. https://doi.org/10.1016/j.ress.2013.06.040
16. Goldbeck N, Angeloudis P, Ochieng WY (2019) Resilience assessment for interdependent urban infrastructure systems using dynamic network flow models. Reliab Eng Syst Saf 188:62–79. https://doi.org/10.1016/j.ress.2019.03.007

17. UK National Cyber Security Centre (2019) Cyber assessment framework. https://www.ncsc.gov.uk/collection/caf. Accessed 16 Dec 2020
18. Castells M (2011) The rise of the network society. Economy, Society, and Culture. Wiley, The Information Age

Disinformation in the Deployment and Mitigation of CBRNE Attacks in the Age of New Social Media (NSM)

J. F. Hassard and S-R. Sellevåg

1 New Social Media as an Impact Multiplier

The use of new social media in spreading disinformation has been highlighted as a potential weapon or at least an *Impact Multiplier* in their use of Chemical, Biological, Radiological, Nuclear and Explosive weaponry (hereafter "CBRNE"). This chapter gives reflections on ways in which social media could alter, and arguably enhance, the way CBRNE warfare may be waged or prevented in the future. New Social Media ('NSM') is described as those content providers operating over the internet and increasingly through mobile devices. In no particular order, these include YouTube, Facebook (and Instagram), Twitter, as well as non-Western channels.

Propaganda and disinformation have been used in overt and covert warfare for millennia; the use of NSM platforms have broadened the scope of threat actors employing chemical and biological weapons or the threat thereof, in order to achieve their goals and the means by which such tools can be prevented, discouraged, or mitigated against. The multifaceted ways this can occur are for simplicity herein called "influence multipliers" *pre ipso facto* or "impact multipliers" *post ipso facto*; central to the discussion is whether these can change the way society and those charged with protecting it must deploy their strategies in the light of the increasing potency of said multipliers.

While social media have undoubtedly been used successfully by terror groups in recruitment, fundraising and in the planning and promulgation of their activities, it is wise to consider the threat more holistically, both in terms of the perpetrators and

J. F. Hassard (✉)
Department of Physics, Imperial College, London SW7 2AZ, UK
e-mail: j.hassard@imperial.ac.uk

S-R. Sellevåg
Norwegian Defence Research Establishment (Forsvarets Forskningsinstitutt (FFI)), P.O. Box 25, 2027 Kjeller, Norway
e-mail: stig-rune.sellevag@ffi.no

P.D.E. Biggins and D. Chana (eds.), *CBRNE: Challenges in the 21st Century*, Advanced Sciences and Technologies for Security Applications,
https://doi.org/10.1007/978-3-031-17374-5_10

in the way media, old and new, can be used to subvert the rule of law. In addition, it is important to emphasise that it is not just terror groups and their acolytes which may use and be influenced by NSM disinformation. If anything, some governments' propaganda tools have been sharpened and enhanced by NSM. During 2016–2021 in the USA, and for many years and especially in the present Ukrainian conflict in Russia and elsewhere it has been shown that organisations and institutions are demonstrably guilty of publicising what has become known as 'fake news', often weaving a semi-coherent narrative across many strands of media, and with sophisticated targeting of key demographics so as to achieve some political goal. In addition, the use of disinformation has extended to natural disasters such as the COVID-19 pandemic, with falsehoods spread by institutions and individuals who, for example, promote the idea of vaccines being part of some supranational plot.[1]

Going forward, it is useful to disentangle the *use* of chemical and biological (CB) weapons[2] and how they are *portrayed* by both the protagonist (or perpetrator) and the victims and potential victims of such weapons. The former has the option of using information or disinformation as a force multiplier by offering continued or even enhanced threat. They may also choose to deny their use. The latter may choose to claim their use in order to incentivise potential allies in combatting the threat, or to even deny their use in order to reduce potential advantages to their enemy, or to avoid damaging morale amongst their people. Unconnected third parties may also choose to claim ownership of the use of CB weapons in order to enhance their standing, fundraising and negotiation potential. Furthermore, it is important to acknowledge that the use and misuse of social media does not only act as an Impact Multiplier for nefarious groups seeking to undermine or damage society. The use of Social Media also impacts[3] on the ability and *modus operandi* of law-enforcement agencies in their fight against terrorism, and the use of CBRNE agents.

This chapter looks at the way terror groups or individuals, and some state actors will seek to provide disinformation through multiple media channels. We look at the use of information and disinformation and how it can be used for the above purposes, and how disinformation can be countered. While there are many examples of this information in historical times, it is important to state at the outset that NSM technologies have changed the scope and potential of disinformation enormously- and also enhanced the means by which it can be countered.

---

[1] The anti-vaccination movement is a case in point, and at the time of writing, December 2020, showing no signs of moderating. Dr Moncef Slaoui, chief science adviser to the US federal effort to speed vaccine development, said he was "very concerned" about scepticism about the vaccine in some circles...."Unfortunately ... there's been a confusion between how thorough and scientific and factual the work that has been done is..." with Twitter and YouTube preferred media channels for conveying that scepticism.

[2] See https://michaelhartmann.org/research-paper/chemical-warfare/ and https://www.sciencehistory.org/distillations/a-brief-history-of-chemical-war.

[3] See, for example, https://www.cambridge.org/core/journals/social-policy-and-society/article/police-use-of-social-media-transformation-or-normalisation/19E5C6727038F01129B51CE4287D0FFB.

2 Changing Face of Media: The Rise, Fall and Rise of "Walled Gardens"

It is useful to recall how social media have evolved in the last few decades and what it is about them which makes them potentially potent in promulgating information or disinformation about CBRNE. In the 1980s, Internet Service Providers ('ISPs') such as AOL and Compuserve, offered controlled information content inside "Walled Gardens" and even "Walled Secret Gardens" in which users experienced access to a largely pre-selected portfolio of websites, and experienced a sometimes nebulous affinity to their ISP and afforded it considerable credibility. In 1989, the World Wide Web was born in CERN [1] and to some extent, broke down the walls. Users go largely where they wish, though it is important to recall that even in 2020, Google, Bing and Ecosia etc., only index a tiny part of the addressable internet (the remaining majority is held in databases or behind paywalls).

In the 1990s, a plethora of browsers[4] further democratized the internet. Consumer-to-Consumer (C2C), Business-to-Consumer (B2C) and Business-to-Business (B2B) enterprises[5] etc. and then Google in 1998, further reinforced the notion that the internet was authoritative and to be trusted with algorithms enhancing better web search, offering new ways of doing business, new advertising revenue models and new markets. Wikipedia in 2001 further enhanced the perception of authority. By 2005, however, communities were developing like YouTube, now owned by Google, and others, generally concentrating in various geographies.

The growth of internet use has been startling:

- August 6, 1991: info.cern.ch—first-ever website on the Internet.
- January 1st, 2019: 4,100,667,287 internet users (up from 3.42 billion in 2016) making 7–10 billion search queries a day worldwide and 1.94 billion websites on the Internet.
- In the third quarter (Q3) 2018 there were 342 million domain names registered across all top-level domains.

But the Internet is not all visible or indeed transparent. In 2015, Eric Schmidt, Google's former CEO, estimated that it consists of roughly five million terabytes of data. Google's web search[6] had indexed about 200 terabytes of this data. That is a lot of data, but that comprises only 0.004 percent of the whole web.

Tor (open-source software enabling users to use the world wide web anonymously) uses a "virtual tunnelling" system that acts like a multi-layered Virtual Private Network. "*The Darkness Online*", a study published [2] in February 2016 by Moore and Rid of King's College London, performed a deep search that found that more than half of the 5,205 "live" sites on Tor were for something illegal. In all, the researchers found 300,000 addresses within the Tor network, comprising roughly

[4] 1993, Mosaic, 1994 Yahoo Directory made a serious attempt to create a comprehensive directory.

[5] Such as eBay in 1995.

[6] See also https://www.wired.com/story/google-tracks-you-privacy/.

205,000 web pages. By contrast, Google reckoned that the open web comprised 30 trillion pages back in March 2013. The dark web may be dark, but it also was tiny in 2016.

However, the social media sites such as YouTube, Twitter, and Instagram, now owned by Facebook, along with its Russian equivalent V Kontakt,[7] and corporate equivalents such as LinkedIn (owned by Microsoft[8]) and others have begun to re-establish the "Walled Gardens" of the 1980s. It is perhaps humanity's need for identity, and affinity to others, which these sites foster and satisfy. Together with the plummeting penetration of traditional 'old media' content providers, such as local newspapers, (against which an individual's own perceptions of local reality can be tested) this has led to an increasing dependence on, and belief in the NSM sites which fill the vacuum. It is not all negative: sites such as LinkedIn allow sometimes expert discussion across a wide range of issues which can greatly improve the user experience. As such, they are potent channels for information and also disinformation, and their role in promulgating CBRNE weapons—and acting as influence and impact multipliers—should not be underestimated.

The effect of the COVID-19 disease has been unremittingly grim, with the present pandemic effectively enhancing and accelerating a global trend in massive reductions in old media. According to the Tow Center,[9] which tracks these closures, huge losses in traditional print outlets has not been matched by local content providers. This "desertification" of information sources at a local level has profound implications. Individuals no longer generally have direct experience of phenomena or claims made, but rely on globally accessible influencers, not all of whom are committed to information integrity and unbiased portrayal of issues.

The use of the internet as a primary source of (dis)information is already reaching half the world's population with mobile platforms increasingly important, particularly in the developing world. According to the trade website, eMarketer [3], nearly 47% of the world's population used the internet at least once a month in 2017 (either via desktop/laptop or mobile device), a 6.1% increase compared with 2016. According to this source, adoption surpassed the halfway mark in 2019, when 50.6% of the worldwide population accessed the internet.

Social media has transmuted from the Walled Gardens of a captive audience to something more democratic and transparent, to the current situation where large parts of it are regressing back to those Walled Gardens wherein users are more likely to be adherents and believers in whatever message the site is promoting. COVID-19-related closures of traditional local, verifiable news outlets have exacerbated that process.

---

[7] VK.com had over 539 m users as of April 2020 and operates in 90 languages; it has been claimed by some that it has some political biases but does not appear to be involved in any aspect of CBRNE.

[8] Acquired in 2016 at $26.2bn with over 700 m users in over 150 countries as of May 2020; LinkedIn also promulgates largely corporate themes but is generally regarded as being effectively self-regulating.

[9] See for example, https://www.cjr.org/business_of_news/the-journalism-crisis-is-gaining-momentum.php.

Neglecting the reach of the internet in the use, prevention, mitigation, and normalization of CBRNE—an area which is reasonably opaque to most internet users who are therefore vulnerable and susceptible to disinformation—would therefore be irresponsible. The lessons for future engagements involving CBRNE are clear, whether we are talking about individual, subnational or national groups, whether it be a "natural" event, such as the COVID-19 pandemic, or a deliberate release of biological warfare agents (BWA) or chemical warfare agents (CWA), the use of NSM must be integral to the response, because it will certainly be integral to the actions of the perpetrator.

While "old media" remains susceptible to disinformation, and that this can be used in what we have classed an "impact multiplier" of CBRNE use, it is with new social media that opportunities arise for all protagonists in the threat and risk space. The internet has the potential for extraordinarily rapid and dispersed cloning—and extremization—of malicious content. The rate of propagation offers huge opportunities for those who may wish to use CBRNE as weapons of terror.

The power of NSM has for instance been reinforced by the Russia Today (RT) TV channel: The editor-in-chief, in justifying an expansion on RT's reach, complains[10] that:

> "There is no objectivity…If the channel [RT.RU] does not have a large audience, then it will never be able to influence any opinion. Including about Russia…These are people who are tired of the mainstream….These are people who understand that the whole truth cannot fit into one concept of the world, promoted by the usual Anglo-Saxon media. These are younger people, because we see how actively we are watched through YouTube - about 1 million views a day. By the way, on YouTube, we consistently overtake all international news channels, including CNN International, Fox News, Sky News, Al Jazeera. We already have over 700 million views - a world record for news media. This makes us happy because our viewers are people who will rule our world tomorrow….

This is not to argue that RT has no right to challenge the "Anglo-Saxon" media. On the contrary, our central thesis is that if directed responsibly, NSM can be regarded as a net positive. However, the role of NSM such as RT (stand-alone or through, for example, YouTube) and the role of algorithms must be considered important in how the West approaches its strategies for CBRNE prevention and mitigation, and in the event of an attack, the response necessary.

Furthermore, it is not to suggest that Twitter, YouTube and Facebook or RT should shut down. However, given their ubiquity and power, it is clear that society must do better in reacting to them—in terms of understanding, speed, potency, reach and effectiveness. Indeed, it has been shown that a society in which social media activity is important can react better to even natural disasters. The 2004 Indian Ocean Tsunami, for example, [4] led to communities which were conversant or expert in social media setting up early warning systems and disaster training programmes. If we apply these lessons to non-natural disasters and consider the speed at which

[10] https://www.kommersant.ru/doc/1911336 (my translation)….

disinformation can spread then we can begin to devise NSM strategies to counter any future CBRNE-related actions.

3 Matryoshka Conspiracies?

When one considers the way some countries use NSM in connection to CBRNE episodes, one has to wonder if the Impact Multiplication relies on a combination of deceit and openness in acknowledging what happened. A Matryoshka doll (often called "Babushka doll" in the West) has several layers of hollow wooden doll, each doll but the largest being enclosed in a similar one which is larger. When considering some recent CBRNE-related events, one is forced to ask: can the perpetrators really have been so incompetent, or is one seeing a smaller, more potent political message wrapped in a larger one?

An example is the alleged perpetrators involved in the Skripal incident in Salisbury who did not apparently realise that they could be tracked by their phone records, something Bellingcat[11] duly did. The Skripals, Sergei and his daughter Yulia, were poisoned in Salisbury with nerve agent Novichok in March 2018; Sergei had worked in Russian Military Intelligence (GRU), and since 1995 had connections[12] to British intelligence: he was swapped in 2010 in the Russian/US Illegals Programs Spy Swap and settled in the UK. The Bellingcat website[13] rapidly identified the two key suspects, who were travelling under fake identities in the Novichok poisoning of Sergei and Yulia Skripal, and the subsequent death of Dawn Sturgess from the discarded perfume bottle containing the agent. What is noticeable is the use of a range of NSM outlets, which in the end did little to disguise the provenance and indeed, acted to reinforce the notion that the GRU was behind the poisoning. Even the choice of poison, Novichok, seems designed to send a message.

A second example is the poisoning of the Russian opposition activist Alexei Navalny. In August 2020, the investigative website Bellingcat [5] claimed that an undercover team working for Russia's Federal Security Service (FSB) poisoned Navalny after shadowing him on multiple previous trips since 2017, soon after he announced he would contest the Russian Presidency. The shadowing part has been confirmed by official Kremlin sources.

---

[11] www.theguardian.com/world/2020/jun/23/skripal-salisbury-poisoning-decline-of-russia-spy-agencies-gru.

[12] See The Skripal Files: Mark Urban, Macmillan, Oct 2018; For a Russian perspective see "Skripal passed information on a corruption scheme involving Patrushev to British intelligence" https://theins.ru/news/120020 (in Russian).

[13] https://www.bellingcat.com/news/uk-and-europe/2019/06/28/the-gru-globetrotters-mission-london/ Bellingcat, an apparently independent investigative website, working with the Russian Insider team (https://theins.ru/, *op cit.*), used a variety of telematic tracking tools and a sophisticated knowledge of new media fingerprints in Russia and elsewhere.

The impact amplification relating to the CWA use is huge, with Twitter a prime medium both in seeking the facts[14] and in Russian attempts to deflect, dismay, dissemble, diffuse and distract.[15] One example of this illustrates the synergies between NSM and more traditional media. The Russian Ambassador to London, held a press conference and claimed the Skripals had been poisoned and kidnapped by the British security service. This was echoed by RT [6] on the 13th of April 2018. This led to a concerted NSM deluge, including a statement from a Russian Foreign Ministry spokeswoman: *"Even if it's Novichok, about 20 countries produced it. About 20 countries could have developed this chemical material—why have they pointed at Russia?"*.

The Skripal denial by the Kremlin was refuted[16] in Twitter by Bellingcat which itself spawned a surge in retweets. As a sign that previous warning by officialdom had no impact, the Russian Ambassador was also summoned to meet the UK Foreign Secretary after the Navalny poisoning. Perhaps more profoundly, these episodes illustrate the point: NSM and its use in espionage is no longer an activity exclusive to governments. As Jonathan Eyal, of RUSI, stated in the New York Times: "Now it belongs to anyone who has the brains, the spunk and the technological ability…We are witnessing a blurring of distinctions."

But can the Skripal/Navalny episodes really be as inept as they appear?[17] It is the contention of this chapter that rather, they illustrate the burgeoning use of NSM to wrap stories in fast-moving layers of obfuscation and deceit. This has profound lessons for any rational response to future CBRNE attacks.

4 Mitigation Strategies

The first strategy for society to adopt is to recognise there is a problem. Central to this is the way algorithms drive many viewers to more and more extreme content. Few constraints exist on the major NSM platforms, and these are largely retroactive. However, the typical NSM metastasises malicious content very rapidly, and a NSM reactive response even just an hour later is already likely to be too late. A further NSM issue is the way first-responders, governments and military agencies respond, based on the available information.

Secondly, society must recognise that CWA and BWA attacks are not that common in the recent past, but by no means unheard of. Indeed, CWAs can be regarded as a "democratised" form of WMD, unlike nuclear or even radiological, as they are so cheap to produce. This 'spreadability' matches the democratisation of

[14] https://www.bellingcat.com/news/uk-and-europe/2020/12/21/if-it-hadnt-been-for-the-prompt-work-of-the-medics-fsb-officer-inadvertently-confesses-murder-plot-to-navalny/.

[15] https://twitter.com/bellingcat/status/1050851469707100160?lang=en.

[16] See, for example, https://twitter.com/bellingcat/status/1050851469707100160?lang=en.

[17] www.theguardian.com/world/2020/jun/23/skripal-salisbury-poisoning-decline-of-russia-spy-agencies-gru.

NSM. Disturbingly, while instances of CWAs in the last few decades have occurred, they have not been of primary concern to Western media, governments or populations. Furthermore, the likelihood of future use of CWAs and BWAs is likely to increase, particularly as we see perpetrators in many countries effectively getting away with their use or threat while international strictures are ignored.

The third point is that the threat from BWAs and CWAs is greater now than ever. Here "threat" encompasses the likelihood of occurrence and its likely impact. That impact is multidimensional: there is potential for BWAs and CWAs to afflict humans, animals or agriculture. The COVID-19 pandemic has reinforced what was already well-known: an engineered contagious BW attack could incapacitate/kill similarly large numbers.

The fourth point is that expertise is available to anybody who can afford it, whether it be to manufacture the chemical and biological agents or to devise and execute the NSM accompanying actions.

The fifth point is that the main NSM platforms are accreting power and influence far faster than people might think.

Lastly, we must accept that there is no insuperable moral, religious, ethical, operational, organisational, financial, technical or scientific barriers to many parties, whether individuals, sub-national groups or governments obtaining CWAs or BWAs or in adding in Impact Multipliers through New Social Media.

It is also important to realise that there is a major difference between r*eaching* an audience and *effecting* a terrorist action: the case of the terrorist attack on two Mosques in Christchurch, New Zealand illustrates the difference between belief and action. According to the ICCT report: [7]

> Sageman and McCauley and Moskalenko[18] note that there are two types of radicalization. The first is gaining extreme beliefs, or what is known as "cognitive radicalization," while the second is a change in behavior which can lead to violence, known as "behavioral radicalization."

[18] Quoted in ICCT report: Marc Sageman, "Misunderstanding Terrorism", University of Pennsylvania Press 2017 and Clark McCauley and Sophia Moskalenko, "Friction: how conflict radicalizes them and us", Oxford University Press 2017.

According to McCauley and Moskalenko:

> individuals with radical ideas are 100 times more common than individuals involved in radical action; targeting ideas rather than actions multiplies the enemy by a factor of a hundred.

Again, there is a match between the potential of NSM in influence and impact multiplication, the lower entry barrier and potential ubiquity of CWAs and the increasing radicalisation enabled and even promoted by NSMs. As is dicussed elsewhere, the algorithms behind NSMs promote a radicalisation process through offering links to ever more radical sites, since this increases user 'stickiness'. Society must continue to allow lawful activities, freedom of expression, but clearly needs to do better to correlate actions which, when undertaken together, lead to terror activities. Society needs to foster trusted sources, and work to keep them as bias-free as possible.

Individuals, NGOs, academics, and governments need to educate themselves on the fast-moving world of the internet. This is crucial at a time of instability and rapid political change: "The idea of an open, borderless internet is gone, and it's not likely to come back," according to Darrell West, a senior technology fellow at the Brookings Institution.[19] "It's just not possible anymore."

The explosive increase in New Social Media has led to their use by those involved in CBRNE activities, as victims, claimed victims or potential victims, by perpetrators, and by those seeking to confuse the populations, governments and officials in whose interests it is to know the truth. NSM acts as an impact multiplier before, during and after any attack or suggestion of an attack. It can act as an influence multiplier for those who wish to promote the notion that they are CWA and BWA capable and ready (for intimidation or dissuasion purposes), or to dissemble and disguise their use.

It is crucial that those charged with defending society take seriously the threat of NSM as impact multipliers and/ or influence multipliers. This is particularly important in the area of CBRNE, or in situations in which CBRNE could be deployed, with a strong emphasis on the 'C' part. This requires a comprehensive strategy both at government level and within society. It requires that those NSM platforms undertake a comprehensive review of their strategies for identifying and dealing with disinformation. Time is of the essence: a misjudged response to an ongoing attack could exacerbate the problem. However, as seen in the Tokyo sarin attacks and elsewhere, lack of timely and accurate first-responder information caused loss of life and an amplification of the attack's impact. This in turn clearly requires a very sophisticated approach to avoid biases.

Furthermore, independent platforms should be encouraged, and the same level of factchecking which presumably occurs in the best newspapers and the best websites be introduced to all those currently promoting memes which can be, and have proven to be, highly dangerous. NSM propagation and its subsequent impacts are so fast that minutes count. This time domain also matches those assocoiated with the development and deployment of CWAs.

---

[19] https://www.bloomberg.com/news/articles/2020-12-23/internet-freedom-will-be-an-uphill-battle-for-joe-biden.

Equally, it is vital that free speech and availability for a platform for someone with a controversial viewpoint should not be curtailed or cancelled. The storming of the Capitol Building in Washington on the 6th of January 2021, if nothing else, provides a powerful dataset for society to test and refine the way NSM platforms impact on society. It is common throughout history that populations seek sources of information which reinforce their existing notions and even prejudices.

Finally, society needs to be able to use the power of NSM and advanced tools such as Machine Learning and correlation analysis to anticipate those who seek to use CBRNE and other weapons. Balancing out intrusiveness by the authorities with all individuals' right to be heard remains the perennial most difficult task. CBRNE raises the stakes for this goal.

References

1. Berners-Lee T (1989) Information management: A proposal.http://www.w3.org/History/1989/proposal.html. Accessed December 2020
2. Moore D, Rid T (2016) Cryptopolitik and the Darknet. Survival 58(1):7–38
3. https://www.emarketer.com/content/social-media-update-q4-2020. Accessed January 2021
4. http://www.homelandsecuritynewswire.com/dr20150709-social-engagement-helps-in-disaster-preparedness Accessed 31st December 2020
5. https://www.bellingcat.com/news/uk-and-europe/2020/10/23/russias-clandestine-chemical-weapons-programme-and-the-grus-unit-21955/ Accessed December 2020
6. https://www.rt.com/uk/424041-alexander-yakovenko-skripal-poison/ Accessed 31st Dec 2020
7. Veilleux-Lepage Y, Daymon C, Amarasingam A (2020). The Christchurch attack report: key takeaways on Tarrant's radicalization and attack planning. International Centre for Counterterrorism (ICCT). The Hague

Looking Forward—Towards a Comprehensive Approach

P. D. E. Biggins, D. S. Chana, and S-R. Sellevåg

1 Introduction

The preceding chapters have illustrated that CBRNE in an uncertain world presents growing challenges for Nation States not only from a defence perspective but increasingly from a security and resilience one. Aspects discussed have illuminated the complexity of CBRNE and highlighted some of the risks associated with the advancement in technology and reliance on the evolving global digital backbone. The increasing integration of digital communication, command and control and information (DC3I) systems through all aspects of our daily lives and our reliance on these systems is creating new vulnerabilities. The linkages and interrelationships are becoming more complex. This growing level of complexity will become more noticeable as populations move to higher density urban areas. The concept of smart cities whilst very attractive can only come into fruition by being reliant on trustworthy DC3I firmware and software. Unfortunately, the growing number of global cyber security crimes coupled with social influence/manipulation via social media platforms are increasing. This amalgam emphasises the growing interconnectedness and complexity of the social infrastructures both physically and digitally, presenting significant challenges to organisations and authorities which are charged with protecting society and the national infrastructure.

P. D. E. Biggins · D. S. Chana
Institute for Security Science and Technology, Imperial College, London SW7 2AZ, UK
e-mail: p.biggins@imperial.ac.uk

D. S. Chana
e-mail: d.chana@imperial.ac.uk

S-R. Sellevåg (✉)
Forsvarets ForskningsinstituttNorwegian Defence Research Establishment (Forsvarets Forskningsinstitutt (FFI)), P.O. Box 25, 2027 Kjeller, Norway
e-mail: stig-rune.sellevag@ffi.no

P.D.E. Biggins and D. Chana (eds.), *CBRNE: Challenges in the 21st Century*, Advanced Sciences and Technologies for Security Applications,
https://doi.org/10.1007/978-3-031-17374-5_11

Whilst aspects of CBRNE have been addressed previously and emerging issues raised, the key strand is that these materials continue to pose significant risks at both a defence and security level. The evolving web of deterrence that is in place through the global agreements such as UNSCR 1540, Global Partnership (GP), Australia Group (AG), Wassenaar Arrangement (WA), Biological and Toxin Weapons Convention (BTWC) and the Chemical Weapons Convention (CWC) provide mechanisms for Nation States to manage and control proliferation of material/materiel. These conventions and agreements have been formulated over time but still require countries to adhere to the principles laid down. The conventions and agreements will continue to evolve as thinking on them develops on how to protect societies. It is only through human behaviour and motivation that the peaceful aspects of CBRNE can be realised. Unfortunately, even with the web of deterrence, our history has shown these materials have been used for malevolent purposes; whether the materials are used for single assassinations (for example Novichok, Ricin and ^{210}Polonium and the postal delivery of anthrax USA) or the use of chemical warfare agents in regional conflicts in recent times (Iraq and Syria).

Of growing concern is that CBRNE knowledge was once restricted in access but is now becoming more and more widespread through the Internet and advancements in technology. To put this comment into perspective it is worth looking at the growth of synthetic biology as an example: access to genetic engineering in terms of knowledge, equipment and expertise was originally limited to a small group of scientists about 40–50 years ago, but this has changed in more recent times. As shown in Fig. 1 knowledge and access to technologies such as gene editing, gene drives, and gene synthesis is now available to a wide range of actors with limited or no oversight from established industry or governmental organizations. This does raise legitimate concerns about potential violations of the Biological Weapons Convention (BWC) as noted in 2018, by the States Parties of the BWC [1]. This example illustrates the challenges that traditional institutional enforcement organisations or authorities face with respect to transformative technologies in CBRNE coupled with DC3I. Whilst this example focuses on synthetic biology, it is also applicable to chemical, radiological, nuclear and explosives.

The stark reality of the COVID-19 pandemic has shown our vulnerability to rapid global disease spread. It has also shown the disparity on how nations have responded and the difficulty in obtaining an understanding of its origin. However, the global digital backbone infrastructure was sufficiently robust to enable communications facilitating the global supply lines to function and for the exchange of information, allowing core businesses within Nations and between Nations to function with effect.

The first global pandemic of the 21st Century has re-enforced the need to act collectively to help each other. The 17 goals set out in 2015 as part of the UN Resolution on—the Sustainable Development Goals (SDG)—"Transforming our world: the 2030 Agenda for Sustainable Development" highlights the need for cooperative interaction. Member states of the UN are committed to a shared vision of "a world free of poverty, hunger, disease and want, where all life can thrive." As such, many of the SDGs (Fig. 2) with their targets provide direction for the governance of CBRNE risks in a multifaceted threat environment. For example, food security is important

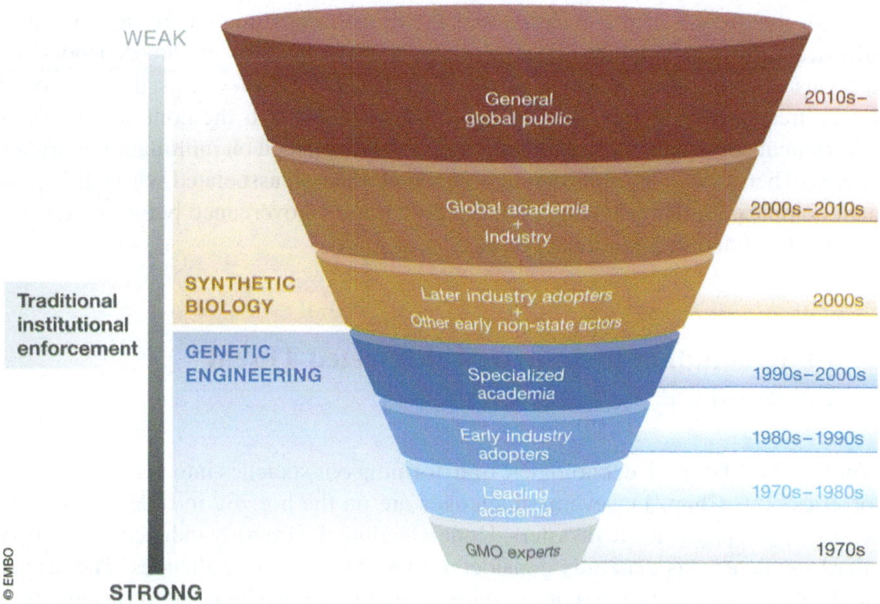

Fig. 1 Increasing number of global users able to access genetic engineering (blue) and synthetic biology (orange) technologies over time [1] (used under the terms of the CC BY 4.0 license)

to end hunger (SDG 2), whilst reducing the number of deaths and illnesses from hazardous chemicals present in air, water and soil and also providing safe, effective and affordable essential medicines and vaccines for ensuring healthy lives (SDG 3). This can only be achieved through inclusive, safe, resilient and sustainable cities and human settlements (SDG 11), including resilient infrastructure (SDG 9).

Fig. 2 The United Nations Sustainable Development Goals for 2030

With the growth of urbanisation and the move to creating smart cities resilient infrastructure is paramount. There will be a growing level of interdependencies in terms of controlling sections of the infrastructure. However, as these cities of the future develop, the delivery will be primarily through the contracting out of government services to private contractors and the reliance on multinational organisations. Therefore, it is important to understand the risk associated with CBRNE in such complex frameworks and examine the issue of governance between Nations, organisations and multinational entities.

2 Understanding Risks in a Multifaceted Threat Environment

With the rise of digital technologies, transforming our societies into so-called "smart societies" (*cf*. Chap. 1), several challenges are on the horizon in order to improve infrastructure resilience to disasters. Understanding the fragility induced by multiple interdependencies is generally considered one of the major challenges. The second challenge is multiple hazards because infrastructures are vulnerable to various stressors, e.g. flooding, earthquake, terrorism and sabotage. Consequently, it is important to find solutions that consider the multiple hazards, including CBRNE hazards, in a comprehensive manner. Lastly, as highlighted by the SDGs, sustainability issues need to be addressed. This covers not only the environmental impact of the disaster on the infrastructure, but also the effects of the infrastructures on the environment, on our climate and on our ecosystem. Thus, in order to understand how modern societies can prepare for and become more resilient towards CBRNE risks, we need to view such risks as part of a multifaceted threat environment.

Characteristic for the multifaceted threat environment is a wide scope of risks ranging from small-scale risks (i.e. only affecting local communities) to slow-onset global catastrophic risks like climate change. This complexity may lead to fragmentation and siloed risk management across sectors (*cf*. Chap. 9), which may create gaps that can render the management of risks ineffective. This implies that a coherent approach across sectors is needed.

The *Sendai Framework for Disaster Risk Reduction 2015–2030* (SFDRR) adopted at the Third UN World Conference in Sendai, Japan, on 18th March 2015 aims to provide such a comprehensive approach. The goal of the SFDRR is to:

> Prevent new and reduce existing disaster risk through the implementation of integrated and inclusive economic, structural, legal, social, health, cultural, educational, environmental, technological, political and institutional measures that prevent and reduce hazard exposure and vulnerability to disaster, increase preparedness for response and recovery, and thus strengthen resilience [2].

However, the SFDRR is not intended to be the primary regulatory instrument. Rather, the framework aims to inform governments and sector authorities on how to implement coherent and holistic risk reduction policies and practices across sectors for a multitude of hazards. To do so, the SFDRR introduced a set of principles and priority areas where the principles are essential to understand what needs to be done in accordance with the priority areas. Of the thirteen principles that were formulated, the following six can be considered to constitute the "backbone" [2]:

- The primary responsibility of states to prevent and reduce disaster risk;
- The shared responsibility between central and local authorities, sectors and stakeholders;
- Protection of persons and their assets, while promoting and protecting all human rights including the right to development;
- Engagement by all of society;
- Full engagement of all state institutions of an executive and legislative nature at national and local levels;
- Coherence of policies, plans, practices and mechanisms across different sectors and agendas.

Clearly, the SFDRR suggest that a whole-of-society approach is needed, requiring empowerment and inclusive, accessible and non-discriminatory participation. Strengthening resilience through effective governance of risks requires, however, a deep understanding and a 360° view of the risks involved. This raises new questions: is such an approach cost-effective? Who has primacy and is it even possible to involve all stakeholders?

Among safety professionals, relating to natural or accidental events, it is common to define risk as a combination of consequences and associated probabilities or associated uncertainties. In security settings, on the other hand, the triplet asset/value, threat and vulnerability is commonly used to define risk, which may indicate that assessment of probability is considered inappropriate for deliberate malicious events. Furthermore, based on the associated disastrous consequences, some risks can be considered as *existential* or *catastrophic* as is the case for nuclear weapons, pandemics and climate change. Other risks are associated with great uncertainty and can be considered as *emerging*, while *systemic* risks may occur in systems with great complexity. In the discussion of how to safeguard societies against natural, accidental or deliberate events, governance of emerging and systemic risks is of great importance.

3 Systemic and Emerging Risks

Our society and critical national infrastructures can in many ways be characterised as complex adaptive systems. Risks that evolve because of the inherent properties and emergent behaviour of complex adaptive systems can be considered as "systemic risks" [3]. Thus, systemic risks appear due to connections between risks. As a result, a localised initial failure can lead to cascading failures due to the strong interactions. Systemic risks can therefore be defined as "threats that individual failures, accidents or disruptions present to a system through the process of contagion" [4]. Characteristic for systemic risks is that such risks span more than one sector of the economy and often span more than one country. Such risks therefore have broad ramifications for our society.

Governance of emerging security risks is of particular concern. An emerging risk is defined by the International Risk Governance Council (IRGC) as a "risk that is new, or a familiar risk in a new or unfamiliar context or under new context conditions (re-emerging)" [5, 6]. The sources of emerging risks can be either natural (e.g. bacteria and viruses), human (e.g. terrorism) or the result of causal interactions between different sources which makes it difficult to designate the risk source as either natural or human, which is often the case for systemic risks [7]. Emerging risks are, per se, affected by significant uncertainty and a lack of knowledge about possible consequences for risk-absorbing systems. This must therefore be taken into consideration when establishing effective management strategies. To aid in understanding the complex interconnected nature of risks it is useful to develop and run scenarios in a "what if" mode.

4 Use of Scenarios

General morphological analysis is a method for structuring and investigating the total set of relationships in multidimensional, non-quantifiable problems [8, 9] that is especially suited for so-called "wicked problems" [10] and scenario generation [11]. The methodology consists of repeated sequences of analysis and synthesis phases. The methodology begins by identifying and defining a set of parameters that frame the problem under study. An exhaustive and mutually excluding set of possible values must be decided for each parameter. The next step is to create the so-called morphological box that contains all possible configurations of the problem space that can be constructed on the basis of the parameter set. However, the morphological box usually contains numerous pairs of mutually incompatible configurations. A cross-consistency assessment to reduce the total set of configurations into a smaller set of internally consistent configurations that represent the solution space is therefore carried out. In this assessment, each pair of configurations is examined, and a judgement is made whether the pair represent a consistent relationship (without any reference to causality or direction). Consistency is assessed on the basis of two criteria: *logical contradictions* and *empirical inconsistencies* (i.e. empirically implausible or highly improbable combinations). As such, the morphological process only determines if a solution is consistent or not; no assessment of any cause-and-effect relations are made. Finally, the solution space is surveyed, and the appropriate solutions are selected.

Scenarios therefore provide guidance to decision-making. However, it is important to realise that scenarios should not be considered as predictions of future events. Rather, scenarios are "tool[s] for ordering one's perceptions about alternative future environments in which one's decisions might be played out" [12]. Furthermore, scenarios must conform to the principles of *plausibility*, *consistency* and *relevance* [13]. This type of approach can aid authorities and organisations to develop appropriate governance of risks.

5 Governance of Risks

Given the difficulties of understanding risks in a multifaceted threat environment, what are the implications for risk governance? As established by the International Risk Governance Council (IRGC), risk governance applies the principles of good governance to the identification, assessment, management and communication of risks. However, what is meant by "governance" and what constitutes "good governance"?

The word "governance" is ubiquitous and used in a variety of contexts. At a general level, it refers to all forms of social coordination and patterns of rule or simply put, the process of decision-making and the process by which decisions are implemented. In order to understand what constitutes "good governance", the eight principles used by the United Nations provide guidance:

Participation
Rule of Law
Consensus Oriented
Equity and Inclusiveness
Effectiveness and Efficiency
Accountability
Transparency
Responsiveness

The principles of "good governance" are applied in the IRGC risk governance framework and can be applied to the governance of risks in a multifaceted threat environment.

6 A Risk Governance Framework

As a starting point to risk governance the IRGC [14, 15] have developed a framework. In the context of this framework, *risk* is defined as "an uncertain consequence of an event or an activity with respect to something that humans value", while *hazards* describe "the potential for harm or other consequences of interest". *Threat*, on the other hand, is defined as "a stated or inferred intention to initiate an attack with the intention to inflict harm, fear, pain or misery". Furthermore, a capability to inflict harm is also needed in order to constitute a real, significant threat. In the same context, it is also of interest to define the term *vulnerability*, which can be considered as "the degree to which a system is affected by a risk source" [16].

The novelty of the IRGC framework is that it integrates scientific, economic, social and cultural aspects. In addition, it includes effective engagement of stakeholders. This makes the framework applicable to a range of different risks. Specifically, the framework is applicable for so-called "simple", "complex", "uncertain" and "ambiguous" risk problems.

The characterisation of a particular risk depends on how difficult it is to establish a cause-effect relationship between a risk agent and its potential consequences, the reliability of the cause-effect relationship and what the risk actually means for those affected and the values to be applied when deciding what to do with the risk. A strategy is then derived for risk assessment, risk management and the stakeholder participation. Consequently, the framework also takes the societal context into consideration, i.e., the structure and interplay of different actors dealing with risks, how these actors may perceive the risks and what concerns they have regarding the consequences of the risks. Furthermore, policies, regulations and socio-political impacts are also taken into consideration.

The risk process of the IRGC framework consists of four main phases: "pre-assessment", "appraisal", "characterisation and evaluation" and "management" (Fig. 3). The purpose of the pre-assessment phase is to frame the problem space, identify and monitor early warning signals of the risk, screen for models, policies or best practices for dealing with the risk and, finally, select major assumptions, conventions and procedural rules for assessing the risk and its associated emotions. Given the CBRNE proliferation concerns that follow the rapid advances in science and technology as well as the risk that non-state actors may acquire and use advanced technologies for crimes or terrorism, proper framing of the problem space and identification and monitoring of early warning signals is important, but also challenging due to high complexity and uncertainty. For this purpose scenario generation is often used. However, the problem becomes even more challenging when considering covert use of CBRNE as a means of hybrid warfare. Understanding the system under study is therefore of utmost importance for proper problem framing.

After the pre-assessment phase, the next phase is risk appraisal. Risk appraisal not only covers risk assessment, but also an assessment of related concerns and social and economic implications. Consequently, risk appraisal is a two-stage process where the purpose of the first stage—the risk assessment—is to produce the best estimate of the physical consequences that a risk may induce. Whilst the purpose of the second stage—the concerns assessment—is to assess the concerns of individuals and of the society as well as calculating the social and economic implications of the risk. The final outcome of the risk assessment will be a specification of the risk together with calculated probabilities for the consequences.

This is not as easy as it sounds. Taking terrorism as an example, estimation of probabilities of such intentional adverse acts is very difficult if not impossible. At best, a judgemental, knowledge-based estimate of the likelihood can be achieved if sufficient intelligence is available. Although more knowledge on terrorism as a phenomenon and better intelligence will aid the risk assessment, it will still be affected by substantial aleatory (stochastic) uncertainty. Furthermore, the risk assessment may also be

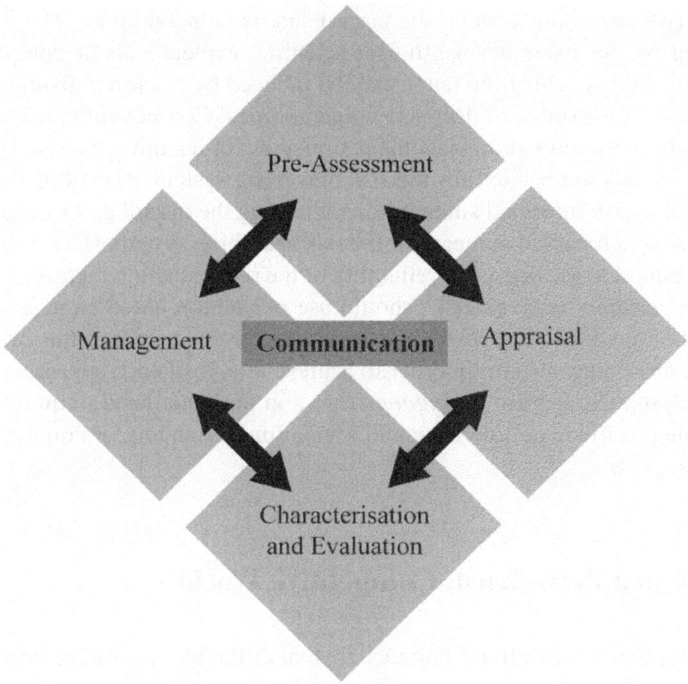

Fig. 3 Risk governance framework (Redrawn S-R. Sellevåg)

affected by substantial complexity (e.g. interdependencies between critical infrastructures). High complexity and uncertainty may also introduce ambiguity to, for example the severity of the threat. Coming back to the principles of "good governance", it is therefore important that the risk assessment is transparent with effective engagement and participation of stakeholders.

The risk characterisation and evaluation phase aims at judging the acceptability and/or tolerability of a risk. This process must not only be informed by the scientific evidence from the appraisal phase and an understanding of risk reduction and mitigation measures, but also an evaluation of broader value-based issues that may influence the judgement. Consequently, the judgement is based on both evidence and values. This is important because we cannot base our judgement on evidence only when deciding what our society is supposed to tolerate or accept. However, this balancing between evidence and values may generate controversial decisions, thus highlighting the importance of risk communication and involvement of stakeholders.

When it comes to risk management, different strategies are needed for the different classes of risks. For *simple* risk problems, traditional risk–risk trade-offs, risk–benefit analysis and cost-effectiveness studies are the methods of choice, as well as best practices. *Complexity-induced* risk problems arise as a result of difficulties in identifying and quantifying causal links between a risk agent (the hazard or threat) and specific observed effects. Complexity may also arise due to insufficient scientific evidence about the risk absorbing system (the target) and its vulnerabilities. The former can be induced by, for example insufficient scientific evidence about complex dose–effect relationships, whilst the latter may be induced by scientific dissent about the effectiveness of measures to decrease vulnerabilities. Consequently, it is necessary to distinguish between risk management strategies for handling the risk agent from the strategies that are needed for the risk absorbing system. The IRGC framework recommends a risk informed strategy for addressing the hazard and its causal chain, and a robustness focussed strategy for the risk absorbing system. Here, robustness is to be understood as the degree of reliability of the risk reduction measures. Management of *uncertainty-induced* risks should use precaution based strategies to avoid the hazard or the threat, and resilience focused strategies to reduce the vulnerability of the risk absorbing system. In order to achieve this, resilience approaches need to be whole system strategies. *Ambiguous* risks, on the other hand, require discourse based strategies to create tolerance and mutual understanding of conflicting views and values.

7 Trust in a Zero-Trust, Competitive World

For wicked problems like hybrid threats, several difficulties are on the horizon when trying to apply the IRGC framework. First of all, if it is difficult to frame the problem in the pre-assessment phase, how should the appraisal, characterisation and evaluation, and management of such risks be carried out? Secondly, if the framing problem is solved, the next problem would be to identify and monitor early warning signals of the risk. This is difficult because traditional intelligence methods using indicator-based methods for early warning of hostile intent and activity are not very applicable to hybrid threats.

During the Cold War, the job of Western intelligence services was to solve puzzles of the type: how many nuclear war heads has the Soviet Union? Adversaries using hybrid threats, on the other hand, shape themselves to our vulnerabilities and exploit the faults in our defences and preparedness measures. Consequently, the risks posed by hybrid threats need to be considered as mysteries, not puzzles that can be solved. Furthermore, where puzzle-solving becomes easier when new information is added, the problem with mystery-solving is that the judgement and assessment of large amounts of uncertain and possibly false information becomes harder. Framing mysteries therefore involves "adjusting subjective probabilities in light of new, but inconclusive, evidence" [17]. Consequently, hybrid threats are risks affected by significant uncertainty.

In a multifaceted threat environment, pre-assessment of risks has therefore moved from solving puzzles to framing mysteries.

However, it is too simplistic to characterise hybrid threats as an uncertainty-induced risk problem. It is very much also a complexity induced risk problem.

Emerging and disruptive technologies not only add complexity to our critical infrastructures giving rise to systemic risks; such technologies are also changing the nature of the threat because technology is a major driver for almost all types of organised and serious crimes, terrorism and national security threats.

Furthermore, economic globalisation and large demographic changes following migration can also be considered as a form of complexity. When the so-called normative order of common authority and shared values is perceived to be threatened, authoritarian predispositions of the populace can become activated. According to Karen Stenner [18], persons lacking openness to experience dislike variety, novelty, diversity and complexity, while those with cognitive limitations will prefer simplicity. Consequently, lack of openness and cognitive incapacity reduce a person's willingness and ability, respectively, to deal with complexity. Such persons may be vulnerable to conspiracy theories (*cf.* Chap. 10).

Throughout history there are numerous examples of the threat posed by conspiracy theories: the Holocaust, the 22 July terror attacks in Norway and the 2021 storming of the US Capitol are all examples of which the horrendous consequences conspiracy theories can lead. When groups in the society feel powerlessness and alienated, this will erode the trust between the population and the government. This gives fertile ground for conspiracy theories. By playing on the same emotions, i.e. powerless and alienation, conspiracy theories can be used as an instrument of power for harming one of the core foundations of liberal democracies, namely trust. When political polarisation is increasing and trust in society is eroding, fragmentation and segregation amongst the population will occur, pushing the society towards a tipping point.

However, as long as the great-powers rivalry is viewed as a zero-sum game, how should hybrid threats be governed in a multifaceted threat environment? If we look back on the last 100 years or so of the history of terrorism, terrorist strategies have consisted of asymmetrical warfare, waging a war of attrition, use of propaganda and charismatic leadership, establishing legitimacy for the cause and provoking an overreaction by the governments [19]. Despite our knowledge of terrorist strategies, governments still meet terrorism with the use of military force, detention and human rights violations, including torture. Such counter-terrorism strategies will only lead to more terrorism. If liberal democracies do not live up to their standards, terrorist organisations can use this for propaganda and building legitimacy for their cause, making arguments about Western hypocracy. Thus, if governments' deterrence and countering of hybrid threats yields human rights violations, this will only make our society more vulnerable to hybrid threats. As pointed out by the scientist András Rácz [20]:

> Hybrid warfare is built on capitalizing on the weaknesses of a country, on flaws in its political system, administration, economy and society. If an adversary cannot detect sufficient weaknesses, then no full-scale attack can be launched, meaning that hybrid warfare never reaches the second, attack phase. Hence, the best defence against hybrid warfare is good governance.

Further [20]:

> [G]ood governance needs to be interpreted in the broad sense. In addition to a democratic political structure and well functioning public administration, it includes respect for human rights, transparency, media freedom, the rule of law and proper rights guaranteed to ethnic, national, religious and other minorities, all in order to improve the domestic democratic legitimacy and support of the government, and hence the very stability of the state.

Coming back to the IRGC framework, three aspects are of utmost importance in order to ensure successful risk governance:

(i) open, transparent and inclusive communication;
(ii) engaging stakeholders to assess and manage risks;
(iii) fully consider the societal context of both the risk and the decision that will be taken.

Effective communication during all phases of the risk governance process is therefore needed in order to establish long-term trust in risk management, especially when risks are perceived as complex, uncertain or ambiguous. This can only happen if there is sufficient trust in the society in the first place. In the multifaceted threat environment, this requires safeguarding universal respect for human rights. It can therefore be argued that ensuring long-lasting trust and respect for human rights are at the core of governance of security-related risks.

8 In Conclusion: Towards a Comprehensive Approach

This book has attempted to provide a modern treatment of CBRNE threats, within the context of a world that is increasingly proliferated with computing, communications, and advanced data processing. When first considering the compilation of this volume—several years ago—the editors had a notion that within certain security circles the consideration of physical threats had somewhat fallen out of fashion, with the topic of *cybersecurity* appearing to many to be a new and separate realm of threat consideration that somehow usurped the need to further consider seemingly dated and passé CBRNE risks. It is our contention that such compartmentalisation was and still is, highly counterproductive from the perspective of predicting, mitigating, and reacting to threat scenarios of the future. This is simply because a coordinated blend of physical and cyber techniques would seem to be a logical strategy for a sophisticated attacker to adopt, given the opportunity to do so. Some researchers and decision makers still choose to confine their work within tightly defined cliques of cybersecurity and physical security, as though one is disassociated with the other. This suggests that a volume such as this is well motivated, if for no other reason than to highlight that such compartmentalisation is, at best, an arrangement of convenience to bound a problem space to avoid it becoming unwieldy and, at worst, ignores the inspection of *the* most important cyber-physical security risks of today and tomorrow. Regretfully, at the time of writing the risk of dogmatic adherence to such threat compartmentalization is all too clearly illustrated by recent global security events including: the

COVID-19 pandemic; the consequences of a disastrous withdrawal of military forces from Afghanistan by NATO allies; and the outbreak of a brutal armed conflict in the country of Ukraine, due to an invasion by Russia. When examined diligently, each of these cases represent security incidents that are characterised by a complex interplay between physical and cyber elements, with the war in Ukraine even being dubbed the 'TikTok war' in recognition of the high importance of cyber factors in what is a very physical conflict. In one way or another, this collection of globally consequential security events spans the full array of CBRNE and cyber factors over a period of just 24 months in the world's recent history; which illustrates the timeliness of the discussions presented here. Amidst the complex interplay of these threat factors and the points made regarding them in preceding chapters, it is instructive to distill some key themes that have emerged when considering the modern CBRNE threat environment and security more generally.

8.1 Uncertainty, Interconnectivity and Risk in CBRNE

As we have seen earlier in this volume, when considering any threat, we ideally aim to have some quantitative understanding of the risk that it might pose through an understanding of two fundamental quantities—the probability of the threat occurring and the impact or severity of the threat/event, were it to occur. Most simply and for the purpose of the discussion in this section, we might derive a value of risk, R, from these two quantities by considering their product such that:

$$R = L \times I$$

where L is some score of probability and I is some score of impact/severity on a fixed scale; such that an impact score of zero would relate to a completely benign event and a maximum score to a complete catastrophe. It should be recalled here that the process of assigning values to both L and I—and R as a result—is often subjective and typically only accessible through some semi-quantitative process of assessment and deliberation. As has been alluded to earlier in this text, there are more sophisticated formulations of risk that can be derived, but for highlighting the core contemporary problems of accurate CBRNE risk quantification, which is our intent, this *simple risk equation* will suit our purposes adequately.

There can be little argument that the level of interconnectivity of 'things' in our world has increased significantly since the emergence and subsequent rapid adoption of the internet and mobile telecommunications in the 1990s. In a few decades, the idea of a powerful pocket computer has transitioned from a concept of science fiction to a somewhat mundane aspect of daily life—at least in the wealthiest parts of the world—and we can project that such a reality will continue to become the norm for increasing numbers of the world's population in the coming years and decades. Such interconnectivity, however, is not just confined to interactions between human beings. It brings with it the opportunity for our cyber-physical network of humans and machines—people, software and hardware—to communicate with each other and, powered by techniques such as machine learning and trends towards greater levels of autonomy, the agency of our machines to make physical and digital changes in the world with minimal human input or interaction. There are many reasons why such developments promise transformative positive change to people across the globe, but since our aim here is to discuss the evolution of security risks, we dwell briefly on this point only to serve as a reminder of the drivers that motivate and compel such technology trends; these trends are likely to be irresistible and so halting their advance should be ruled out of our set of viable security risk mitigation options. When examined through a security lens, interconnectivity presents some growing challenges with respect to our current methods and processes for assessing and, subsequently, managing threats and risks. The increased complexity that interconnectivity brings, leads to a greater uncertainty in our determination of CBRNE threat probabilities and also limits our ability to adequately quantify the impact of a given security scenario. Add to this the compounding factor of disinformation and the ability to confuse populations at scale using automated information generation systems that utilise the power of algorithms to target certain messages to certain demographics and we can see that well known CBRNE threats take on layers of complexity that are seldom considered in conventional analyses. Furthermore, this increased complexity brings with it another significant issue of the emergence of *novel* threats, e.g. through the composition and coordination of attacks using well understood threat elements or by the emergence and exploitation of new vulnerabilities in our societal machinery. These features of complexity and interconnection motivate us to consider security threats on ever widening geographic scales where, increasingly, there are direct causal connections between citizens and the various components of a global socio-technical system, where previously there may have been none. The current trends of isolationism (at the time of writing) in many political spheres around the world would appear to run counter to this wider-context security model that our technology motivates, serving to increase such risks, rather than diminish them. Therefore, it should not come as any surprise that global response systems to the pandemic seemed so flat-footed and compartmentalised around the world, nor that the global financial system is struggling to deal with a systemic level economic event that the war between Russia and Ukraine has precipitated; with a global split between those who vocally condemn a perceived act of aggression from Russia and those that view it as nothing more than a similar catastrophe to ones they may have experienced in their geographic locale. In both cases, an inability to inspect CBRNE factors in holistic terms has led to significant loss of life and long-term damage. For those who read the preceding sentences and feel that the issues drawn out are far too nuanced or, more typically, 'only loosely connected' to CBRNE concerns—such views and perspectives are precisely the problem.

8.2 *The Big N, CBRNE Rhetoric and Technology Literacy*

In recent years we have seen the emergence of emboldened public rhetoric on the use of nuclear weapons from the presidents of both the world's leading nuclear weapon states—the USA and Russia. The Intermediate- Nuclear Forces Treaty has been abandoned by both and, perhaps most worryingly, the Non-Proliferation Treaty itself has been undermined by actions such as the UK's unilateral decision to reverse its policy of nuclear arms stockpile reduction and transparency, in 2021. Within 80 years of their use in World War II, it seems that 'nuclear options' for warfare are back on the table due, rather ironically, to a perception of increased global instability and insecurity from rapidly developing economic powers like China. In 2022 it is still the case that major geopolitical power stems from the ownership of nuclear arsenals; with the decision-making structure of the five permanent members of the United Nations security council still in place and the threat of nuclear weapons deployment being used as both passive and active threats in multiple security flashpoints around the world. Although an entire industry of qualitative analysis exists around the topic of nuclear weapons and CBRNE, a simple and elegant examination of this situation, stemming from our deliberations on risk, is that normalising the concept of nuclear weapons use, increases both the probability of their use and their potential impact. A world that tacitly agrees that order and power should be maintained by the ultimate expression of CBRNE disruptive force, is likely to encourage the further development and use of CBRNE threats as a mechanism to exert power, coerce and determine conflict.

Furthermore, the limitations of governments and decision makers to understand the science of CBRNE issues and the technology landscape that society is building and is being shaped by, undermines confidence that strategists are capable to adequately gauge the significance of CBRNE threats and are able to identify and track viable methods for managing such risks. A world where some feel it is credible to opine about the techniques for countering a biohazard by ingesting bleach, virus propagation by mobile-phone networks, the use of chemical warfare agents, the validity of climate change research, the feasibility of using 'tactical' nuclear weapons, the scale of the threat posed by quantum computers and the weaponising of machine learning, without basic levels of technical literacy, represents our current lived reality.

8.3 *Outlook for the Future*

In the coming decades, a series of rapidly emerging and disruptive technologies (EDTs), will continue, if not accelerate, the pace of change that digital hardware and software technologies have already delivered in the last half century. Harnessing the quantum scale, for example, is likely to result in changes in how we undertake computations, how we develop materials, understand biological and chemical processes and how we transmit and store information. Space-borne systems will increasingly form a part of our collective global critical physical infrastructure and our digital spaces are set to transition from mere recreational or information technology environments to a real extension of human existence, through the innovations that are being delivered by crypto-currencies, blockchains, digital twins, Web 3.0 technologies and the concept of metaverses. The trends of decentralisation in such metaverse environments mean that existing industries and governance mechanisms are being disintermediated, leading to a stronger sense of web community that potentially changes notions of nationality, citizenship and security. Managing the development of CBRNE threats and risks within such environments will present numerous unique challenges. One could argue that the digital means employed by non-nation state organisation such as ISIS represent an early indication of how transnational CBRNE threats are likely to evolve in the future.

Although this book has focused predominantly on ways in which CBRNE tactics and systems might be used to cause harm, it is worth noting that the science and technology underpinning many of the domains that have been touched upon here also represents the means by which we might most sustainably address the security risks that have been our focus. In this chapter we mentioned the United Nations Sustainable Development Goals (SDGs), whose solutions present a pathway for sustainable global security. If directed correctly, CBRNE research and development has the potential to provide significant contributions solving many of the global challenges that drive conflict today and are likely to continue to drive conflict tomorrow, including:

- The treatment of disease and advancement of medicine
- Solving issues of food and water scarcity
- Improving infrastructure efficiency and resilience
- Addressing climate change
- Solving the humanity's energy budget problem, e.g. through nuclear fusion

Appropriately guiding efforts towards productive CBRNE science and technology, in preference to destructive uses, whilst managing persistent and evolving CBRNE threats and risks will require strategic vision and commitment. Adequate technical literacy within national institutions and other global decision-making entities represents a key factor in achieving the necessary strategic competence needed to maintain this balance and is, therefore, the most critical aspect of CBRNE security that needs maintenance, particularly given the trajectory of cyber-physical and societal complexities that have been covered by our contributing authors in this volume. With the growth of a global technology economy, it is likely that the competition for technically literate talent will become even more challenging for institutions, such as governments, suggesting that the situation is likely to deteriorate unless steps are taken to proactively correct it. It is our hope that this text will contribute to inspiring, edifying, and motivating students to understand a CBRNE security landscape that with which it is vital for us all to be engaged.

References

1. Trump BD, Galaitsi SE, Appleton E, Bleijs DA, Florin M-V, Gollihar JD, Hamilton RA, Kuiken T, Lentzos F, Mampuys R, Merad M, Novossiolova T, Oye K, Perkins E, Garcia-Reyero N, Rhodes C, Linkov I (2020) Building biosecurity for synthetic biology. Molecul Syst Biol 16(7):e9723. https://doi.org/10.15252/msb.20209723
2. UNISDR (2015) Reading the Sendai framework for disaster risk reduction 2015–2030. https://www.preventionweb.net/files/46694_readingsendaiframeworkfordisasterri.pdf
3. International Risk Governance Council (2018) Guidelines for the governance of systemic risks
4. Centeno MA, Nag M, Patterson TS, Shaver A, Windawi AJ (2015) The emergence of global systemic risks. Ann Rev Sociol 41:65–85
5. International Risk Governance Council (2010) The emergence of risks
6. International Risk Governance Council (2015) Guidelines for emerging risk governance
7. International Risk Governance Council (2010) Emerging risks: sources, drivers and governance issues
8. Zwicky F (1969) Discovery, invention, research through the morphological approach. The Macmillan Company
9. Ritchey T (2013). General morphological analysis. A general method for non-quantified modelling. Swedish Morphological Soc. https://www.swemorph.com/pdf/gma.pdf. Accessed 2 April 2021
10. Rittel H, Webber M (1973) Dilemmas in a general theory of planning. Policy Sci 4:155–169
11. Ritchey T (2013) Wicked problems. Modelling social messes with morphological analysis. Acta Morphologica Generalis 2:1–7
12. Schwartz P (1996) The art of the long view. Planning for the future in an uncertain world. Doubleday
13. van der Heijden K (2005) Scenarios. The art of strategic conversation (2nd ed.). Wiley
14. International Risk Governance Council (2005) Risk governance: towards an integrative approach (White Paper No. 1)
15. International Risk Governance Council (2017) Introduction to the IRGC risk governance framework (Revised Version 2017)
16. Aven T, Ben-Haim Y, Andersen HB, Cox T, Droguett EL, Greenberg M, Guikema S, Kröger W, Renn O, Thompson KM, Zio E (2018) Society for risk analysis glossary. https://sra.org/sites/default/files/pdf/SRA%20Glossary%20-%20FINAL.pdf
17. Treverton GF (2007) Risks and riddles. Smithsonian Mag. https://www.smithsonianmag.com/history/risks-and-riddles-154744750/
18. Stenner K (2020) Authoritarianism and the future of liberal democracy. HOPE Not Hate Mag. https://www.hopenothate.org.uk/2020/11/01/authoritarianism/
19. Parker T (2014) Avoiding the terrorist trap. Why respect for human rights is the key to defeating terrorism (Vol 12). https://doi.org/10.1142/p995 . World Scientific
20. Rácz A (2015) Russia's hybrid war in Ukraine (FIIA Report 43). The Finnish Institute of International Affairs. https://stratcomcoe.org/andras-racz-russias-hybrid-war-ukraine-breaking-enemys-ability-resist